KT-437-422

Introductory Robotics

J. M. Selig
*Department of Electrical and Electronic Engineering,
South Bank Polytechnic*

NORWICH CITY COLLEGE LIBRARY		
Stock No.	119283	
Class	629.892 SEL	
Cat.		Proc.	

Prentice Hall

NEW YORK LONDON TORONTO SYDNEY TOKYO SINGAPORE

119 283

First published 1992 by
Prentice Hall International (UK) Ltd
66 Wood Lane End, Hemel Hempstead
Hertfordshire HP2 4RG
A division of
Simon & Schuster International Group

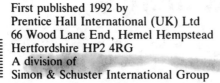

© Prentice Hall International (UK) Ltd, 1992

All rights reserved. No part of this publication may be
reproduced, stored in a retrieval system, or transmitted,
in any form, or by any means, electronic, mechanical,
photocopying, recording or otherwise, without prior
permission in writing, from the publisher.
For permission within the United States of America
contact Prentice Hall Inc., Englewood Cliffs, NJ 07632.

Printed and bound in Great Britain by Dotesios Ltd,
Trowbridge, Wiltshire

Library of Congress Cataloging-in-Publication Data

Selig, J. M.
  Introductory robotics/J. M. Selig
    p.   cm.
  Includes index.
  ISBN 0–13–488875–8 (pbk.)
  1. Robotics.   I. Title.
  TJ211.S434 1991
  629.8'92—dc20   91–23704
                                                       CIP

British Library Cataloguing-in-Publication Data

Selig, J. M.
  Introductory robotics.
  I. Title
  629.8

  ISBN 0–13–488875–8

---

1 2 3 4   95 94 93 92

To my mother and the memory of my father

Contents

Preface

This book grew out of a third year optional course taught to electrical engineering students at South Bank Polytechnic. A parallel course on robot dynamics and control was taught by a colleague. For completeness, I have added here my own treatment of robot dynamics. The control of robots is, however, a very large subject area, which really requires a book of its own. Many such texts already exist.

The scope of this book forms a consistent whole. The kinematics and dynamics of robots are the essential basics on which all of current industrial robotics is built. At present no book concentrates on this material. Certainly all basic texts in robotics mention these subjects, but often only in a cursory manner. The treatment is invariably in terms of many co-ordinate frames and festooned with indices. One of the main purposes of this book is to present the kinematics of robots in as simple and clear a manner as possible. This involves only using one co-ordinate frame and then using active rather than passive transformations to describe the positions of rigid bodies.

Using a simple notation we are able to study real six joint robots. However, we approach these examples via planar and spherical devices. These small examples are also interesting in their own right as they are often component parts of larger machines. The inverse kinematics is often difficult for students, usually because they have never had to solve systems of non-linear equations. Before attacking these problems, the difficulties which can arise in such systems are introduced.

The study of manipulator jacobians is sometimes called infinitesimal kinematics. We look at a selection of the applications of the jacobian, but the main applications are in the statics and dynamics of robots which are discussed separately. Six component vectors called instantaneous screws are introduced. Also a simple proof is given that the columns of the jacobian are the robot's joint screws. Once again we build up from planar and spherical examples.

Trajectory planning is usually associated with problems of obstacle avoidance. Trajectory following by contrast is really an application numerical analysis; in particular, function interpolation. We must approximate curves in joint space which correspond to desired movements of the robot's end-effector.

The study of forces and torques in static equilibrium is a precursor to robot dynamics.

But it also has some useful applications in its own right, especially to gripping solid objects. A different type of six component vector, called a wrench, is introduced. The difference between screws and wrenches is that they transform differently under rigid body transformations. A pairing between wrenches and screws is also introduced; it gives the work done by a wrench acting on a screw. In classical screw theory, the work done is given by the reciprocal product of two screws. Following such a style would make inertias more difficult to handle later. Also, the introduction of wrenches, with different transformation properties from screws, is more in keeping with modern approaches to classical mechanics. The introduction of wrenches allows a straightforward derivation of the force on the end-effector due to the joint torques. It is also an important theoretical tool for rigid body dynamics.

To keep the exposition of robot dynamics simple, we use wrenches and screws. Using wrenches we get a single six-vector equation for each rigid body. The dynamics of robots are derived by writing the Newton-Euler equation for each link and projecting out the joint torque component. This does involve the introduction of some more 'machinery', namely the vector product of screws and the product of screws and wrenches. The same equations can be derived using Lagrangian mechanics; however, it was assumed that students would only be familiar with Newtonian mechanics.

The only other assumptions made are a knowledge of trigonometry, matrix algebra and calculus, to the level of first year science or engineering degree. In fact it is hoped that a mature 'A'-level maths student could cope with most of the material in the book.

Throughout the book no specific units have been used, except that angles are always measured in radians. This is not intended to mean that units are not important, just that it does not matter if we use metric or imperial units. Most American robots have lengths quoted in inches and weights in pounds, while Japanese and European models use millimetres and kilograms.

It is a cliché that research and teaching support each other. For me, writing this book and teaching the course it was based on have been a practical demonstration. I could not have contemplated teaching such a course without my previous research interest. As a result of the research I felt that it was possible to simplify the presentation of robotics. Then, in the course of the teaching I was forced to solve many problems. Usually the problems were old ones, but the challenge was to present the solution as clearly as possible and in a manner consistent with the rest of the course. This has very definitely helped my research.

I would like to thank colleagues at South Bank and elsewhere for their inspiration and encouragement, in particular S. Adams, C.G. Gibson, A. Kaposi and T. Sattar. Finally I apologize to past students who suffered the course as it developed.

London 1991

1   Introduction

1.1   What is Robotics?

Many books and articles on robotics begin with a dictionary definition of a robot. In general, it is a good idea to define our terms at the beginning, and be explicit as to what we are talking about. Unfortunately, there is no commonly agreed definition of a robot, and several well known dictionaries contain no definition at all. Most attempts at a definition say something like 'a robot is a machine that is capable of being reprogrammed'. The fact that a robot can be reprogrammed is important: it is definitely a characteristic of robots. But of course, the definition given above is much too crude. It includes computers and microprocessor controlled washing machines, which we do not usually think of as robots. The trouble is that if we try to be more specific about what kind of machines robots are, we will exclude some robots. For example, if we insist that a robot must be capable of manipulating solid objects, we ignore all sorts of paint spraying and welding robots. Suppose we said that robots have to be in the form of an arm with some kind of tool or end-effector in the place of the hand: then we would leave out the walking robots under development in several places. Like all technology, robots perform tasks which are idealizations or extensions of human capabilities. So we cannot use the anthropomorphic character of robots to define them.

The difficulty stems from the fact that ideally we would like robots to be general purpose machines, capable of almost anything. So we do not want to place limits on what robots can achieve or how they are constructed. At present, though, robots can do very little, so it is hard to distinguish them from machines with only limited capabilities. For the present our definition will be similar to the well known definition of an elephant: 'difficult to describe but you will know one if you see one'.

Within its short history robotics has become an extremely large and eclectic discipline. The study of robots involves elements of both mechanical and electrical engineering, as well as control theory, computing and now artificial intelligence. Underpinning much robotics is the mathematics of rigid body transformations. This is the main subject area of the present book. We will look at the simple theory of moving solid objects around in space using current six joint industrial robots. However, the theory also applies to parallel manipulators like the Stewart platform, the fingers of multi-finger hands and the legs of walking machines. The book only aims to cover a small part of the subject; the basic kinematics and dynamics of robots. The details of motors to drive the robot's joints, sensors and other technical concerns are very important, but will be left to others to describe.

1.2 Popular Robotics

Mention robots to most people and they usually think of 'mechanical' beings, human in every respect except their (re)production. Most of us have our favourite robot from films or TV, Marvin the Paranoid Android from *The Hitch Hiker's Guide to the Galaxy* perhaps. Older readers may recall 'Robby the Robot' from *Forbidden Planet* or the robot from *Lost in Space*; even older ones 'Maria' from Fritz Lang's *Metropolis*.

There must be some reason why we have such a fascination with robots. One of the deepest characteristics of humans is their talent for creating tools and machines. Yet we have not succeeded in creating a living being. If we were to make such a creature, would it be conscious? This is still a matter of controversy within the philosophy and artificial intelligence communities. Some scholars believe that anything that behaves like a conscious being, is a conscious being. The argument might be caricatured as 'if it walks like a duck and quacks like a duck, it is a duck'. Other scholars are not convinced. In the popular conception robots have always been conscious beings and not the simple automata currently used in industry.

The word robot comes from the Slavic word meaning work or worker. It came into the English language via the play *Rossum's Universal Robots* by the Czech author Karel Capek. The plot, as with many early robot stories, harks back to Mary Wollstonecraft-Shelley's great gothic novel *Frankenstein*. A man creates a living being and in the end it destroys him. Perhaps this is supposed to suggest that people are playing God and get no more than they deserve. A more sinister interpretation might be that we are destined to kill, or have already killed, our creator.

In its turn *Frankenstein* harks back to older folk tales and legends: the Gollem, for example, a clay figure brought to life by a rabbi in Prague; also the ancient Greek legend of Pygmalion, a statue brought to life. It is interesting to note how these stories reflect the knowledge of their time: the older stories use magic and religion, while *Frankenstein* uses the then emerging biological sciences, especially the experiments by Galvani on frogs' legs.

In the 1950s the mood of robot stories changed, probably exemplified by Isaac Asimov's *I Robot* stories. With the new technological age, robots were now perfect machines created in factories. They obey Asimov's three laws of robotics: a simple code of ethics, saying, for example that robots must not kill humans or injure themselves. The fiction then flows from putting the unfortunate robots in situations which produce dilemmas within their simple moral code. Incidentally, the word 'robotics', meaning the study of robots, seems to have been coined by Asimov.

In the late 1960s and early 1970s fiction took a turn against robots as it did against science and technology in general. Several films of this period contain 'bad robot' characters. Most well known of these are probably *Westworld*, *Alien* and *Blade Runner*. If one allows the intelligent computer 'HAL' as a robot, then *2001 a Space Odyssey* would be in this category, as would *Dark Star* where the intelligent bomb could not make up its mind whether or not to explode. Also at this time, other television programmes and films portrayed robots as the unthinking, unfeeling henchmen of the bad guy: ideal for the good guys to blast to atoms, but with no messy blood to frighten the children. The television series *Buck Rogers in the 25th Century* was particularly murderous towards robots, even though 'Tweaky' was the hero's pet.

The 1970s saw a rehabilitation of robots. Now robots turned up as the hero's cute sidekick, for example 'R2D2' and 'C3PO' in *Star Wars*. However, one of the first films like this was *Silent Running* where the maintenance robots were named after Donald Duck's nephews. Modern Japanese culture seems particularly fond of robots. This can be seen in children's cartoons which nearly always contain robot characters.

For someone making a serious study of robotics it is important to remember that the popular conception of robots comes from this cultural tradition and that robots are always thought of as human-like machines. What is more, these ideas predate the modern industrial robots discussed in this book. So in a sense, they are the real robots and it is us who have stolen the name.

1.3 History of the Technology

One evening in the mid 1960s Joe Engleberger met George Devol at a cocktail party. Devol had just patented a design for a computer controlled mechanical arm. Engleberger saw this invention as a primitive robot. The two of them went into business; they formed the company Unimation. The ancestors of the original arm were the telechirs or tele-operated arms developed in the 1940s. These 'master-slave' arms are used to manipulate radio-active materials from behind the safety of lead-glass screens. The human 'master' was replaced by computer control, using technology developed for computer controlled machine tools at M.I.T. Boston. In the early 1970s Engleberger visited Japan, and it was the Japanese who first realized the potential of industrial robots. At the time Japanese industry was investing heavily in new manufacturing plant. Unimation robots were at first manufactured under licence in Japan, then improved models were developed and these spawned Japanese robot companies.

Originally robots were intended to replace human workers. This was especially true in the automotive industry which was suffering from industrial relations problems at the time. It was thought that robots would be universal machines, capable of rapid reprogramming for a wide variety of tasks. It was these ideals which motivated the development of the Puma by Unimation in a research project for General Motors. It was closely modelled on the human arm and could lift about the same weight.

In practice the early robots were hard to reprogram and could not compete with humans in tasks where the location of the work pieces was not precisely known. So the first applications for these new machines was paint spraying and welding. Here the robot could be programmed by a human operator leading it through the required sequence of movements. Progress was made in assembly tasks with development of the Scara robot. This was developed in Japan, for mounting components on printed circuit boards for the electronics industry.

Around this time the ideology of robotics changed: no longer were robots to replace human workers, but were advertised as being able to do jobs humans could not, such as working in the hazardous environments of the mining, offshore oil and nuclear industries or in fire fighting. Robots can work in places inaccessible to humans, in outer space, on the sea bed or inside contorted pipe work. Finally, robots can work on a scale humans find difficult, for instance in very large scale assembly or handling tasks. Alternatively, robots can manipulate objects at almost microscopic scales.

With the world economic depression of the early 1980s many people saw the adoption of new technology as a possible cure. However, this led to fear of even more unemployment as robots took people's jobs. This does not seem to have happened to any extent and indeed the robot industry has gone through something of a depression of its own in the late 1980s. This may have been due to the fact that robots were sold as a universal panacea for manufacturing industry: no one could have a modern production line without robots, they could do anything. Clearly no technology could live up to such promises.

Moreover, the economic benefits of installing robots came under scrutiny and were found wanting. Certainly, robots do not need holidays and do not go on strike, but they are only machines. And like all machines they need regular maintenance and they will break down. Installing robots in a factory is not simply a matter of replacing workers with robots. The production line will have to be redesigned, the production process itself may have to be changed. Often the product will have to be modified so that robots can manipulate the components easily. Getting rid of unskilled workers involves employing highly skilled robot technicians and programmers; fewer of them to be sure. But the shortage of such personnel has, in some cases, hampered the installation of robots. The problems of applying robots in real manufacturing situations deserves a book in itself, and indeed such books do exist.

1.4   Looking Ahead

Current research in robotics tends to be in two main areas, artificial intelligence and the related field of machine vision. A typical artificial intelligence problem in robotics might be to find a clear path for a robot through a cluttered environment. However, most of the research in this area is towards making intelligent machines. This is pure artificial intelligence and has little to do with robotics, except that any successes would have immediate applications to robots.

Machine vision is another field which might be considered as separate from, but closely connected with, robotics. Getting an electronic representation of a picture or scene into a computer is relatively simple. Using a camera, points in the picture can be encoded according to their light intensity and colour. Interpreting such a representation so that the objects in the scene can be identified is extraordinarily difficult. The archetypical application of such technology to robotics is the 'bin picking' problem. The vision system looks at a bin of disordered components. It must recognize the parts and relay the position and orientation of the correct part to the robot, in such a way that the robot can pick it up.

More generally, much of the current research can be seen as leading to robots capable of working in disordered or cluttered environments and able to deal with greater levels of uncertainty. To achieve this robots need more sensors and of a greater sophistication. Such sensors would certainly include vision, but also touch, force and possibly ultrasonic rangefinding. This leads to the problem of 'sensor fusion'. Not only must the sensor data from several different sources be interpreted, but also conflicts within the data must be resolved. For example, if one sensor says that the wall is 10 cm away but another says it is only 0.5 cm away, which one do you believe? Again these problems are not restricted to robotics.

In the United Kingdom research in these areas has been termed 'advanced robotics', and many developments are underway, particularly in domestic robots, which would perform housework, and medical robots for surgery. Public acceptance of such machines will be a problem, but it may not be too long before such robots are produced. Manipulators which can replace lost human limbs have been possible for some years now. People have been unwilling to use such devices for esthetic reasons. The increasing sophistication of these protheses will probably soon overcome people's reluctance though. The development of robots for use by the disabled is also a small but lively area of research; more esoterically, in Australia a sheep shearing robot has been demonstrated.

There is, however, another strand to current research, and these projects are closer to the mechanical aspects of the subject. There is a lot of interest in redundant manipulators, that is, manipulators with more that six joints. For such manipulators the position and orientation of the gripper can be held fixed while the rest of the arm moves. Most humans can just about do this with their arms. The advantage of such a robot is that it can reach around obstacles and still perform its work. However, programming such a machine is very difficult.

To make robots quicker it is necessary to make the links lighter. This means they will be more flexible. In this book we assume the links are perfectly rigid, an approximation of course. There is much current work on robots with significant flexibility. This is particularly important for robots in outer space, where weight is at a premium. It is also important to reduce the vibrations caused by this elasticity for accurate work on earth.

Mobile robots are also under development in many places. This type of work has quite a long history, dating back to a legged truck built in the 1950s by General Electric. The advantage of using jointed legs rather than wheels or tracks is that legged vehicles may be able to cope with much rougher terrain. Certainly ants manage to walk over extremely rough ground. Another idea borrowed from the insect kingdom is wall climbing robots. Some of the prototype machines look very like spiders; however, they cling to the wall with suction pads or magnets. These robots are intended to have applications in the construction industry.

Multi-fingered dextrous hands are another development based on a biological model. It is a formidable technological task to build such a gripper with a size and complexity comparable to the human hand. Automatically controlling such a device seems to be an even greater challenge. A similar problem under active investigation is how to control two or more co-operating robot arms. It is possible to plan trajectories for both arms so that they do not collide, but doing this quickly enough is a challenge.

Lastly, although most researchers assume that there is no more to know about the serial six joint arm described in this book, a few problems still remain, most notably hybrid control, where both the position and force applied by the robot must be controlled stably.

In conclusion, it is always difficult to predict the future. The breakthroughs which will shape robotics in the next century may not be in any of the above mentioned areas. The above should only be taken as a rather biased attempt to summarize current research in robotics. A final word of warning; readers should be wary of the extravagant claims made by the robotics community. Often machines advertised as general purpose only work in special circumstances. Unfortunately, robotics seems to suffer more than most disciplines from being oversold or hyped.

2   Rigid Transformations

Robots move solid objects around in space. We will assume all solid objects are rigid bodies, including the parts that make up the robot itself. In most cases this is a pretty good approximation, but it can fail. For example, imagine using a robot to grasp a plastic bottle. There is also growing research interest in fast robots made with light and therefore flexible members, particularly for use in outer space. However, we will stick to the rigid body model. What we need is a neat way of describing the possible positions of a rigid body. This can be done by studying the possible rigid body transformations, since if we fix a standard or 'home' position for the body we can describe any subsequent position by giving the transformation needed to get there from home.

Rigid body transformations are characterized by the fact that they preserve the distance between points. Suppose we have two points in the rigid body with position vectors $\mathbf{v}_1$ and $\mathbf{v}_2$. The square of the distance between these points is given by:

$$(\mathbf{v}_1 - \mathbf{v}_2) \cdot (\mathbf{v}_1 - \mathbf{v}_2) = (\mathbf{v}_1 - \mathbf{v}_2)^T (\mathbf{v}_1 - \mathbf{v}_2)$$

After a transformation of the body the points will have new positions, say, $\mathbf{v}'_1$ and $\mathbf{v}'_2$. The new points will be separated by a distance:

$$(\mathbf{v}'_1 - \mathbf{v}'_2) \cdot (\mathbf{v}'_1 - \mathbf{v}'_2) = (\mathbf{v}'_1 - \mathbf{v}'_2)^T (\mathbf{v}'_1 - \mathbf{v}'_2)$$

So if:

$$(\mathbf{v}_1 - \mathbf{v}_2) \cdot (\mathbf{v}_1 - \mathbf{v}_2) = (\mathbf{v}'_1 - \mathbf{v}'_2) \cdot (\mathbf{v}'_1 - \mathbf{v}'_2)$$

for every pair of points $\mathbf{v}_1$ and $\mathbf{v}_2$, the transformation is rigid.

We begin by looking at 2-D or planar transformations.

2.1   Rotations and Translations in 2-D

Rotations are rigid transformations, they do not distort the size or shape of the body. Rotations about the origin in 2-D can be represented by 2×2 matrices of the form:-

$$\mathbf{R}(\theta) = \begin{pmatrix} \cos\theta & -\sin\theta \\ \sin\theta & \cos\theta \end{pmatrix}$$

Notice that we use the notation $\mathbf{R}(\theta)$ for the rotation matrix.

To see why this is, consider the effect of an anticlockwise rotation by θ radians on a general position vector, see fig. 2.1. Assume that $\mathbf{v} = (x, y)^T$ is the position vector of some point on the rigid body. We can put this in polar form by writing $x = r\cos\phi$ and $y = \sin\phi$. After the rotation, the point will have the same r value, but the angle from the x-axis will be $\theta + \phi$. So after the rotation the point will have position vector:-

$$\mathbf{v}' = \begin{pmatrix} r\cos(\theta+\phi) \\ r\sin(\theta+\phi) \end{pmatrix} = \begin{pmatrix} r(\cos\theta\cos\phi - \sin\theta\sin\phi) \\ r(\sin\theta\cos\phi + \cos\theta\sin\phi) \end{pmatrix} = \begin{pmatrix} x\cos\theta - y\sin\theta \\ x\sin\theta + y\cos\theta \end{pmatrix}$$

Using the standard trigonometric identities:-

$$\cos(A+B) = \cos A \cos B - \sin A \sin B$$
$$\sin(A+B) = \sin A \cos B + \cos A \sin B$$

This can be summarized by matrix multiplication by the rotation matrix:-

$$\mathbf{v}' = \mathbf{R}(\theta)\mathbf{v} = \begin{pmatrix} \cos\theta & -\sin\theta \\ \sin\theta & \cos\theta \end{pmatrix} \begin{pmatrix} x \\ y \end{pmatrix} = \begin{pmatrix} x\cos\theta - y\sin\theta \\ x\sin\theta + y\cos\theta \end{pmatrix}$$

There are several things to note about these rotation matrices.

- Clearly a rotation of 0 radians will have no effect, and indeed substitution of $\theta = 0$ into the rotation matrix yields the identity matrix.

$$\mathbf{R}(0) = \begin{pmatrix} 1 & 0 \\ 0 & 1 \end{pmatrix}$$

- The effect of two successive rotations is given by matrix multiplication.

$$\mathbf{v}'' = \mathbf{R}(\theta_2)\mathbf{v}' = \mathbf{R}(\theta_2)\mathbf{R}(\theta_1)\mathbf{v}$$

It is not too hard to see that the combined rotation will be through an angle $\theta_1 + \theta_2$. This can be checked by performing the multiplication:-

$$
\begin{aligned}
\mathbf{R}(\theta_2)\mathbf{R}(\theta_1) &= \begin{pmatrix} \cos\theta_2 & -\sin\theta_2 \\ \sin\theta_2 & \cos\theta_2 \end{pmatrix} \begin{pmatrix} \cos\theta_1 & -\sin\theta_1 \\ \sin\theta_1 & \cos\theta_1 \end{pmatrix} \\
&= \begin{pmatrix} \cos\theta_2\cos\theta_1 - \sin\theta_2\sin\theta_1 & \sin\theta_2\cos\theta_1 - \cos\theta_2\sin\theta_1 \\ \sin\theta_2\cos\theta_1 + \cos\theta_2\sin\theta_1 & \cos\theta_2\cos\theta_1 - \sin\theta_2\sin\theta_1 \end{pmatrix} \\
&= \begin{pmatrix} \cos(\theta_1+\theta_2) & -\sin(\theta_1+\theta_2) \\ \sin(\theta_1+\theta_2) & \cos(\theta_1+\theta_2) \end{pmatrix} = \mathbf{R}(\theta_2 + \theta_1)
\end{aligned}
$$

again using standard trigonometric identities.

- The inverse of a rotation matrix can now be seen to be just a rotation in the opposite direction.

$$\mathbf{R}(\theta)^{-1} = \mathbf{R}(-\theta)$$

By considering the symmetry of the sine and cosine functions we can also see that $\mathbf{R}(\theta)^{-1} = \mathbf{R}(\theta)^T$, the transpose of the matrix. Hence, we have that:-

$$\mathbf{R}(\theta)^T \mathbf{R}(\theta) = \mathbf{I}$$

So if we have two vectors $\mathbf{v}_1$ and $\mathbf{v}_2$, rotating them both does not affect their scalar product, since:-

$$
\begin{aligned}
\left(\mathbf{R}(\theta)\mathbf{v}_1\right) \cdot \left(\mathbf{R}(\theta)\mathbf{v}_2\right) &= \left(\mathbf{R}(\theta)\mathbf{v}_1\right)^T \left(\mathbf{R}(\theta)\mathbf{v}_2\right) \\
&= \mathbf{v}_1^T \mathbf{R}(\theta)^T \mathbf{R}(\theta)\mathbf{v}_2 = \mathbf{v}_1^T \mathbf{v}_2 \\
&= \mathbf{v}_1 \cdot \mathbf{v}_2
\end{aligned}
$$

This justifies our earlier assertion that these rotations are rigid body transformations, since we can apply it to the square of the distance between two points; $(\mathbf{v}_1 - \mathbf{v}_2) \cdot (\mathbf{v}_1 - \mathbf{v}_2)$.

- The determinant of a rotation matrix is given by:-

$$\det \mathbf{R}(\theta) = \begin{vmatrix} \cos\theta & -\sin\theta \\ \sin\theta & \cos\theta \end{vmatrix} = \cos^2\theta + \sin^2\theta = 1$$

This means that a rotation will leave the area of a rigid body unchanged: this is characteristic of all rigid body transformations. The fact that the determinant is $+1$ and not -1 means that the transformation involves no reflections. Reflections are also rigid body transformations but no real machine can effect such a transformation, so we exclude consideration of them. Strictly, we should talk of proper rigid transformations if we exclude reflections.

Next we look at the translations. A translation may be represented by a vector. The effect of a translation on a point with position vector $\mathbf{v}$ is simply to add the translation vector to it; see fig. 2.1. If the translation vector is $\mathbf{t} = (t_x, t_y)^T$, then symbolically the translation will be given by:-

$$\mathbf{v}' = \mathbf{v} + \mathbf{t} = \begin{pmatrix} x \\ y \end{pmatrix} + \begin{pmatrix} t_x \\ t_y \end{pmatrix}$$

Translations are rigid transformations as they preserve the distance between points. Two points, $\mathbf{v}_1$ and $\mathbf{v}_2$, become $\mathbf{v}_1 + \mathbf{t}$ and $\mathbf{v}_2 + \mathbf{t}$. Hence the vector between them is unchanged:-

$$(\mathbf{v}_1 + \mathbf{t}) - (\mathbf{v}_2 + \mathbf{t}) = \mathbf{v}_1 - \mathbf{v}_2$$

The effect of two successive translations $\mathbf{t}_1$ and $\mathbf{t}_2$, will be given by vector addition; $\mathbf{t}_1 + \mathbf{t}_2$.

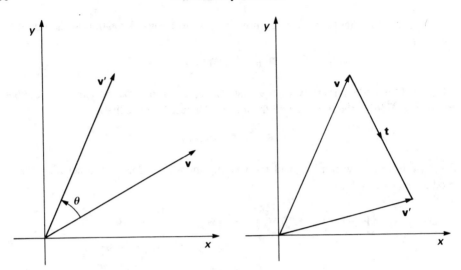

Figure 2.1 Rotation and Translation in the Plane

2.2 General Rigid Motions in 2-D

So far we have not defined what we mean by a rigid body transformation. In fact it is simply a transformation which leaves the distance between any pair of points fixed. It is not too hard to show that any rigid body motion can be broken down into a sequence of translations, rotations about the origin and reflections. As mentioned already we will ignore the reflections and consider only proper rigid body transformations.

A general proper rigid body transformation is given by a pair $\left(\mathbf{R}(\theta), \mathbf{t}\right)$, where, as before, $\mathbf{R}(\theta)$ is a 2×2 rotation matrix and $\mathbf{t}$ a translation vector. These pairs have the following effect on the position vectors of points:-

$$\left(\mathbf{R}(\theta), \mathbf{t}\right) : \mathbf{v} \longmapsto \mathbf{R}(\theta)\mathbf{v} + \mathbf{t}$$

So two successive transformations will give:-

$$\left(\mathbf{R}(\theta_2), \mathbf{t}_2\right) : \mathbf{R}(\theta_1)\mathbf{v} + \mathbf{t}_1 \longmapsto \mathbf{R}(\theta_2)\mathbf{R}(\theta_1)\mathbf{v} + \mathbf{R}(\theta_2)\mathbf{t}_1 + \mathbf{t}_2$$

This is equivalent to the single proper rigid transformation given by the pair:-

$$\left(\mathbf{R}(\theta_2)\mathbf{R}(\theta_1), \mathbf{R}(\theta_2)\mathbf{t}_1 + \mathbf{t}_2\right)$$

Thus we get a sort of 'funny' multiplication rule for these pairs:-

$$\left(\mathbf{R}(\theta_2), \mathbf{t}_2\right)\left(\mathbf{R}(\theta_1), \mathbf{t}_1\right) = \left(\mathbf{R}(\theta_2)\mathbf{R}(\theta_1), \mathbf{R}(\theta_2)\mathbf{t}_1 + \mathbf{t}_2\right)$$

Notice here that pure translations are represented by pairs of the form $(\mathbf{I}, \mathbf{t})$, where $\mathbf{I}$ is the identity matrix. Pure rotations, that is rotations about the origin, are given by pairs

$\left(\mathbf{R}(\theta),\mathbf{0}\right)$. The identity transformation, that is the one which leaves all vectors unchanged, corresponds to the pair $(\mathbf{I},\mathbf{0})$.

The order that these transformations are performed in is very important. That is, the 'funny' multiplication is not commutative. For example, the result of rotating and then translating is different from translating first and then rotating. In the first case we have:-

$$\left(\mathbf{I},\mathbf{t}\right)\left(\mathbf{R}(\theta),\mathbf{0}\right) = \left(\mathbf{R}(\theta),\mathbf{t}\right)$$

whereas the other way round gives:-

$$\left(\mathbf{R}(\theta),\mathbf{0}\right)\left(\mathbf{I},\mathbf{t}\right) = \left(\mathbf{R}(\theta),\mathbf{R}(\theta)\mathbf{t}\right)$$

We may reverse the effect of any transformation by applying the inverse transformation:-

$$\left(\mathbf{R}(\theta),\mathbf{t}\right)^{-1} = \left(\mathbf{R}(\theta)^{T},-\mathbf{R}(\theta)^{T}\mathbf{t}\right)$$

It is an easy matter to check that using the 'funny' multiplication we get:-

$$\left(\mathbf{R}(\theta),\mathbf{t}\right)\left(\mathbf{R}(\theta),\mathbf{t}\right)^{-1} = \left(\mathbf{R}(\theta),\mathbf{t}\right)^{-1}\left(\mathbf{R}(\theta),\mathbf{t}\right) = \left(\mathbf{I},\mathbf{0}\right)$$

The above notation is very cumbersome and it would be much more useful if we could represent the transformations by single matrices. This can be done using the following trick. Consider the 3×3 matrices of the form:-

$$\begin{pmatrix} \cos\theta & -\sin\theta & t_x \\ \sin\theta & \cos\theta & t_y \\ 0 & 0 & 1 \end{pmatrix} \quad \text{or in partitioned form} \quad \left(\begin{array}{c|c} \mathbf{R}(\theta) & \mathbf{t} \\ \hline 0 & 1 \end{array}\right)$$

Multiplication of two of these matrices gives:-

$$\begin{pmatrix} \cos\theta_2 & -\sin\theta_2 & t_{2x} \\ \sin\theta_2 & \cos\theta_2 & t_{2y} \\ 0 & 0 & 1 \end{pmatrix} \begin{pmatrix} \cos\theta_1 & -\sin\theta_1 & t_{1x} \\ \sin\theta_1 & \cos\theta_1 & t_{1y} \\ 0 & 0 & 1 \end{pmatrix} =$$

$$\begin{pmatrix} \cos(\theta_2+\theta_1) & \sin(\theta_2+\theta_1) & t_{1x}\cos\theta_2 - t_{1y}\sin\theta_2 + t_{2x} \\ \sin(\theta_2+\theta_1) & \cos(\theta_2+\theta_1) & t_{1x}\sin\theta_2 + t_{1y}\cos\theta_2 + t_{2y} \\ 0 & 0 & 1 \end{pmatrix}$$

Symbolically, using partitioned matrices this can be written:-

$$\left(\begin{array}{c|c} \mathbf{R}(\theta_2) & \mathbf{t}_2 \\ \hline 0 & 1 \end{array}\right) \left(\begin{array}{c|c} \mathbf{R}(\theta_1) & \mathbf{t}_1 \\ \hline 0 & 1 \end{array}\right) = \left(\begin{array}{c|c} \mathbf{R}(\theta_2+\theta_1) & \mathbf{R}(\theta_2)\mathbf{t}_1 + \mathbf{t}_2 \\ \hline 0 & 1 \end{array}\right)$$

This exactly mimics the 'funny' multiplication of the pairs derived above. We may also use these matrices to find the effect of transformations on the position vectors of points. Suppose $\mathbf{v} = (x,y)^T$ is the vector under consideration. We can turn this into a three-vector by appending an extra 1; $\mathbf{v} = (x,y,1)^T$. This extended vector will also be called $\mathbf{v}$. The

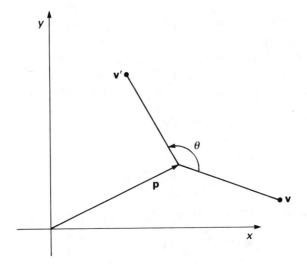

Figure 2.2 Rotation about an Arbitrary Point

extra 1 here is just a mathematical device and **not** the z-co-ordinate of anything. So now we can write:-

$$\begin{pmatrix} \cos\theta & -\sin\theta & t_x \\ \sin\theta & \cos\theta & t_y \\ 0 & 0 & 1 \end{pmatrix} \begin{pmatrix} x \\ y \\ 1 \end{pmatrix} = \begin{pmatrix} x\cos\theta - y\sin\theta + t_x \\ x\sin\theta + y\cos\theta + t_y \\ 1 \end{pmatrix}$$

Again the partitioned form gives a clearer idea of what is going on:-

$$\left(\begin{array}{c|c} \mathbf{R}\,(\theta) & \mathbf{t} \\ \hline 0 & 1 \end{array}\right) \left(\begin{array}{c} \mathbf{v} \\ \hline 1 \end{array}\right) = \left(\begin{array}{c} \mathbf{R}\,(\theta)\mathbf{v} + \mathbf{t} \\ \hline 1 \end{array}\right)$$

This give the co-ordinates of the new point with one appended. This matrix representation is so much more convenient that we will not use the pairs at all.

2.3   Centres of Rotation

So far we have only considered rotations about the origin. Rotations about an arbitrary point in the plane are also rigid transformations. Thus, we ought to be able to find the matrix corresponding to a rotation of θ radians, about some point $\mathbf{p} = (p_x, p_y)^T$ not necessarily the origin, see fig. 2.2.

The motion can be broken down into three stages:

- First translate the point p to the origin; the corresponding matrix is $\left(\begin{array}{c|c} \mathbf{I} & -\mathbf{p} \\ \hline 0 & 1 \end{array}\right)$

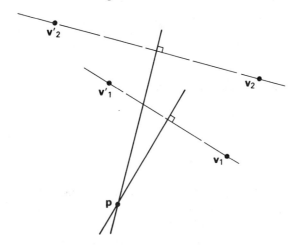

Figure 2.3 Finding the Centre of a Transformation

- Now rotate about the origin, using $\left(\begin{array}{c|c} \mathbf{R}(\theta) & \mathbf{0} \\ \hline 0 & 1 \end{array}\right)$

- Finally, translate the origin back to $\mathbf{p}$ with $\left(\begin{array}{c|c} \mathbf{I} & \mathbf{p} \\ \hline 0 & 1 \end{array}\right)$

Putting all of this together gives:

$$\left(\begin{array}{c|c} \mathbf{I} & \mathbf{p} \\ \hline 0 & 1 \end{array}\right) \left(\begin{array}{c|c} \mathbf{R}(\theta) & \mathbf{0} \\ \hline 0 & 1 \end{array}\right) \left(\begin{array}{c|c} \mathbf{I} & -\mathbf{p} \\ \hline 0 & 1 \end{array}\right) = \left(\begin{array}{c|c} \mathbf{R}(\theta) & \mathbf{p} - \mathbf{R}(\theta)\mathbf{p} \\ \hline 0 & 1 \end{array}\right)$$

Note that the first operation goes on the right. This is easily remembered by considering the effect of the transformation on some point in the plane. Since the matrix representing the transformation is written on the left of the vector representing the point, successive transformation are written further to the left.

This method of finding a transformation by turning it into one that we already know is very useful and we will see it many times. In general an operation on a matrix $\mathbf{M}$, with the shape $\mathbf{T}\,\mathbf{M}\,\mathbf{T}^{-1}$, is called a **conjugation**. In this case we end up with a rotation and a translation, a rotation about the origin by θ, followed by a translation by $\mathbf{p} - \mathbf{R}(\theta)\mathbf{p} = (\mathbf{I} - \mathbf{R}(\theta))\mathbf{p}$. In fact, apart from pure translations, any rigid transformation is of this form, that is a rotation about some fixed centre.

It is possible to find the centre of rotation of a transformation graphically. All we need is two points and their transforms; see fig. 2.3. We must join each point to its transform with a line segment. The perpendicular bisectors of the segments will meet at the centre of rotation. This works because the perpendicular bisectors must be normal to the circular paths taken by the points, and hence be diameters of these circles.

For greater accuracy we need an algebraic technique for finding the centre of rotation.

To do this, compare a general rigid transformation with the one just derived:-

$$\left(\begin{array}{c|c} \mathbf{R}(\theta) & \mathbf{t} \\ \hline 0 & 1 \end{array}\right) = \left(\begin{array}{c|c} \mathbf{R}(\theta) & \mathbf{p} - \mathbf{R}(\theta)\mathbf{p} \\ \hline 0 & 1 \end{array}\right)$$

In order for this equality to hold we must have:-

$$\mathbf{p} - \mathbf{R}(\theta)\mathbf{p} = \mathbf{t}$$

This is a linear equation for $\mathbf{p}$, the centre of the rotation. It can be solved so long as $\theta \neq 0$ and we have:-

$$\mathbf{p} = (\mathbf{I} - \mathbf{R})^{-1}\mathbf{t}$$

Equivalently, the centre of rotation is the point that is fixed under the transformation, that is the point which is in the same position after the transformation. Hence, it must correspond to an eigenvector of the matrix; the one associated with the eigenvalue 1:-

$$\left(\begin{array}{c|c} \mathbf{R}(\theta) & \mathbf{t} \\ \hline 0 & 1 \end{array}\right)\left(\begin{array}{c} \mathbf{p} \\ 1 \end{array}\right) = \left(\begin{array}{c} \mathbf{p} \\ 1 \end{array}\right)$$

This just gives the same equation as above.

To summarize, we have shown that all proper rigid body transformations can be represented by 3×3 matrices of the form:-

$$\left(\begin{array}{c|c} \mathbf{R}(\theta) & \mathbf{t} \\ \hline 0 & 1 \end{array}\right)$$

This can be interpreted as a rotation of θ radians about the origin, followed by a translation $\mathbf{t}$. Alternatively, we may think of the transformation as a rotation about some point in the plane. This point $\mathbf{p}$, called the centre of rotation, can be found by solving the linear matrix equation:-

$$\left(\begin{array}{c|c} \mathbf{R}(\theta) & \mathbf{t} \\ \hline 0 & 1 \end{array}\right)\left(\begin{array}{c} \mathbf{p} \\ 1 \end{array}\right) = \left(\begin{array}{c} \mathbf{p} \\ 1 \end{array}\right)$$

Exceptionally, these equations cannot be solved: this happens when there is no rotation. These transformations as referred to as pure translations, and some people like to think of them as rotations about points 'at infinity'.

Exercises

2.1   Find the 3×3 matrices which describe the following motions in 2-D:

(i)   A $\frac{\pi}{3}$ rotation about the origin.

(ii)   A translation of one unit in the x-direction followed by a $\frac{\pi}{3}$ rotation about the origin.

(iii)   A $\frac{\pi}{3}$ rotation about the point $x = 1$, $y = 1$.

2.2   Find the centre of rotation of the following 2-D motions:

(i)
$$\begin{pmatrix} \frac{1}{\sqrt{2}} & \frac{-1}{\sqrt{2}} & 1 - \frac{1}{\sqrt{2}} \\ \frac{1}{\sqrt{2}} & \frac{1}{\sqrt{2}} & \frac{-1}{\sqrt{2}} \\ 0 & 0 & 1 \end{pmatrix}$$

(ii)
$$\begin{pmatrix} \frac{1}{\sqrt{2}} & \frac{-1}{\sqrt{2}} & 2 - \frac{1}{\sqrt{2}} \\ \frac{1}{\sqrt{2}} & \frac{1}{\sqrt{2}} & 1 - \frac{3}{\sqrt{2}} \\ 0 & 0 & 1 \end{pmatrix}$$

(iii)
$$\begin{pmatrix} \frac{1}{2} & \frac{-\sqrt{3}}{2} & \frac{1+\sqrt{3}}{2} \\ \frac{\sqrt{3}}{2} & \frac{1}{2} & \frac{1-\sqrt{3}}{2} \\ 0 & 0 & 1 \end{pmatrix}$$

2.3 A 2-D rigid motion takes the points $(0, 1)$ and $(1, 1)$ to $\left(\frac{1-\sqrt{3}}{2}, \frac{1-\sqrt{3}}{2} \right)$ and $\left(\frac{2-\sqrt{3}}{2}, \frac{1}{2} \right)$ respectively. Find the 3×3 matrix which effects this transformation and also find its centre of rotation.

2.4 Rotations about the Origin in 3-D

Next we turn our attention to transformations in three dimensions. For convenience, we will use the standard notation in which $\mathbf{i}$, $\mathbf{j}$ and $\mathbf{k}$ represent the unit vectors in the x, y and z directions respectively.

In 3-D any rotation is about some fixed axis. So for a 3-D rotation we must specify the angle of rotation ϕ, as well as a unit vector along the rotation axis, $\hat{\mathbf{v}}$, say. To signify a 3×3 rotation matrix we will write $\mathbf{R}(\phi, \hat{\mathbf{v}})$. It is simple to write down a few such matrices because of our previous work on 2-D rotations. We have:-

$$\mathbf{R}(\phi, \mathbf{k}) = \begin{pmatrix} \cos\phi & -\sin\phi & 0 \\ \sin\phi & \cos\phi & 0 \\ 0 & 0 & 1 \end{pmatrix}$$

The effect of such a rotation on a general point with co-ordinates (x, y, z) is given by:-

$$\begin{pmatrix} x\cos\phi - y\sin\phi \\ x\sin\phi + y\cos\phi \\ z \end{pmatrix} = \begin{pmatrix} \cos\phi & -\sin\phi & 0 \\ \sin\phi & \cos\phi & 0 \\ 0 & 0 & 1 \end{pmatrix} \begin{pmatrix} x \\ y \\ z \end{pmatrix}$$

This shows that the z-component of such a point is always fixed. Hence, the z-axis is fixed, and this is just a rotation in the xy plane.

Similarly we also have simple expressions for rotations in the yz and zx planes:-

$$\mathbf{R}(\phi, \mathbf{i}) = \begin{pmatrix} 1 & 0 & 0 \\ 0 & \cos\phi & -\sin\phi \\ 0 & \sin\phi & \cos\phi \end{pmatrix} \qquad \mathbf{R}(\phi, \mathbf{j}) = \begin{pmatrix} \cos\phi & 0 & \sin\phi \\ 0 & 1 & 0 \\ -\sin\phi & 0 & \cos\phi \end{pmatrix}$$

Notice that the sign of the sine terms are reversed in the last case. This is because $\mathbf{R}(\phi, \hat{\mathbf{v}})$ is, by convention, a rotation of ϕ radians, measured anticlockwise when looking along $\hat{\mathbf{v}}$.

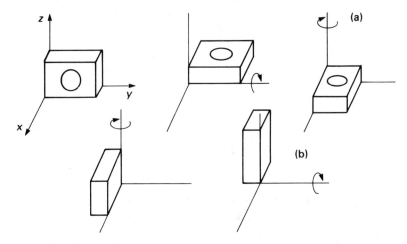

Figure 2.4 Rotations do not Commute

Hence, we must limit the possible values of ϕ (usually we take $0 \le \phi < \pi$), so a rotation of $-\phi$ is represented by $\mathbf{R}(\phi, -\hat{\mathbf{v}})$.

As usual the result of two rotations performed one after the other is obtained by matrix multiplication. This leads us to a very important point: **3-D rotations do not commute**. The order in which we perform the multiplications is important. To illustrate this consider the following rotations:-

$$\mathbf{R}\left(\frac{\pi}{2},\mathbf{k}\right) = \begin{pmatrix} 0 & -1 & 0 \\ 1 & 0 & 0 \\ 0 & 0 & 1 \end{pmatrix} \quad \text{and} \quad \mathbf{R}\left(\frac{\pi}{2},\mathbf{j}\right) = \begin{pmatrix} 0 & 0 & 1 \\ 0 & 1 & 0 \\ -1 & 0 & 0 \end{pmatrix}$$

Let us look at the two possible orders for combining these rotations. To fix our attention think of the effect the rotations will have on a matchbox, see fig. 2.4.

The difference is reflected in the matrix multiplication. In the first case, fig. 2.4(a), we get:-

$$\mathbf{R}\left(\frac{\pi}{2},\mathbf{k}\right)\mathbf{R}\left(\frac{\pi}{2},\mathbf{j}\right) = \begin{pmatrix} 0 & -1 & 0 \\ 1 & 0 & 0 \\ 0 & 0 & 1 \end{pmatrix}\begin{pmatrix} 0 & 0 & 1 \\ 0 & 1 & 0 \\ -1 & 0 & 0 \end{pmatrix} = \begin{pmatrix} 0 & -1 & 0 \\ 0 & 0 & 1 \\ -1 & 0 & 0 \end{pmatrix}$$

Notice that the first operation goes on the right. The other order, fig. 2.4(b), gives:-

$$\mathbf{R}\left(\frac{\pi}{2},\mathbf{j}\right)\mathbf{R}\left(\frac{\pi}{2},\mathbf{k}\right) = \begin{pmatrix} 0 & 0 & 1 \\ 0 & 1 & 0 \\ -1 & 0 & 0 \end{pmatrix}\begin{pmatrix} 0 & -1 & 0 \\ 1 & 0 & 0 \\ 0 & 0 & 1 \end{pmatrix} = \begin{pmatrix} 0 & 0 & 1 \\ 1 & 0 & 0 \\ 0 & 1 & 0 \end{pmatrix}$$

The two answers are certainly different. Also notice that the result is not a rotation about any of the co-ordinate axes.

2.5   General 3-D Rotations

Having seen above some particular rotation matrices, the following question arises: what does a general rotation matrix look like? This is not so easy to answer, however. Indeed, all rotation matrices satisfy the basic relations:-

$$\mathbf{R}(\phi, \hat{\mathbf{v}})^T \mathbf{R}(\phi, \hat{\mathbf{v}}) = \mathbf{I}, \qquad \det \mathbf{R} = 1$$

since rotations must preserve the lengths of position vectors and do not contain reflections.

To find the matrix representing a rotation about some arbitrary vector we may use conjugation. For example, suppose that $\hat{\mathbf{w}}$ is a unit vector in the $x - z$ plane, making an angle of θ with the z-axis, see fig. 2.5. Now a rotation of ϕ radians about this vector can be found by rotating $\hat{\mathbf{w}}$ into coincidence with the z-axis, rotating ϕ about the z-axis and then rotating back to the starting point:-

$$\mathbf{R}(\phi, \hat{\mathbf{w}}) = \mathbf{R}(\theta, \mathbf{j})\mathbf{R}(\phi, \mathbf{k})\mathbf{R}^{-1}(\theta, \mathbf{j})$$

In 2-D it is possible to write any rotation matrix in terms of a single parameter θ, say. Then any rotation matrix would be of the form:-

$$\mathbf{R}(\theta) = \begin{pmatrix} \cos\theta & -\sin\theta \\ \sin\theta & \cos\theta \end{pmatrix}$$

In 3-D it turns out that we need three parameters, but it is impossible to choose the parameters in an unambiguous way. For topological reasons there will always be some choice of parameters which give the same matrix. These imperfect 'local' parameterizations can be useful though; for example, any rotation can be thought of as a product of three rotations about the co-ordinate axes:-

$$\mathbf{R}(\phi_x, \phi_y, \phi_z) =$$
$$\begin{pmatrix} \cos\phi_z & -\sin\phi_z & 0 \\ \sin\phi_z & \cos\phi_z & 0 \\ 0 & 0 & 1 \end{pmatrix} \begin{pmatrix} \cos\phi_y & 0 & \sin\phi_y \\ 0 & 1 & 0 \\ -\sin\phi_y & 0 & \cos\phi_y \end{pmatrix} \begin{pmatrix} 1 & 0 & 0 \\ 0 & \cos\phi_x & -\sin\phi_x \\ 0 & \sin\phi_x & \cos\phi_x \end{pmatrix} =$$
$$\begin{pmatrix} \cos\phi_y\cos\phi_z & \sin\phi_x\sin\phi_y\cos\phi_z - \cos\phi_x\sin\phi_z & \cos\phi_x\sin\phi_y\cos\phi_z + \sin\phi_x\sin\phi_z \\ \cos\phi_y\sin\phi_z & \sin\phi_x\sin\phi_y\sin\phi_z + \cos\phi_x\cos\phi_z & \cos\phi_x\sin\phi_y\sin\phi_z - \sin\phi_x\cos\phi_z \\ -\sin\phi_y & \sin\phi_x\cos\phi_y & \cos\phi_x\cos\phi_y \end{pmatrix}$$

Apart from the fact that this looks terrible, we run into problems when $\phi_y = \frac{\pi}{2}$. The matrix then becomes:-

$$\mathbf{R}\left(\phi_x, \frac{\pi}{2}, \phi_z\right) = \begin{pmatrix} 0 & \sin(\phi_x - \phi_z) & \cos(\phi_x - \phi_z) \\ 0 & \cos(\phi_x - \phi_z) & -\sin(\phi_x - \phi_z) \\ -1 & 0 & 0 \end{pmatrix}$$

So we get the same result whenever $\phi_x = \phi_z + c$, for any constant c.

Another common, local parameterization is in terms of Euler angles. The Euler angles (ϕ, θ, ψ) define the following rotations; rotate ϕ about the z-axis, then rotate θ about the y-axis, and finally rotate ψ about the z-axis again, see fig. 2.5:-

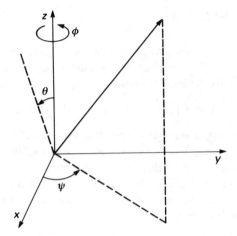

Figure 2.5 Euler Angles

$$\mathbf{R}(\phi, \theta, \psi) = \mathbf{R}(\psi, \mathbf{k})\mathbf{R}(\theta, \mathbf{j})\mathbf{R}(\phi, \mathbf{k})$$

$$= \begin{pmatrix} \cos\psi & -\sin\psi & 0 \\ \sin\psi & \cos\psi & 0 \\ 0 & 0 & 1 \end{pmatrix} \begin{pmatrix} \cos\theta & 0 & \sin\theta \\ 0 & 1 & 0 \\ -\sin\theta & 0 & \cos\theta \end{pmatrix} \begin{pmatrix} \cos\phi & -\sin\phi & 0 \\ \sin\phi & \cos\phi & 0 \\ 0 & 0 & 1 \end{pmatrix}$$

$$= \begin{pmatrix} \cos\psi\cos\theta\cos\phi - \sin\psi\sin\phi & -\cos\psi\cos\theta\sin\phi - \sin\psi\cos\phi & \cos\psi\sin\theta \\ \sin\psi\cos\theta\cos\phi + \cos\psi\sin\phi & -\sin\psi\cos\theta\sin\phi + \cos\psi\cos\phi & \sin\psi\sin\theta \\ -\sin\theta\cos\phi & \sin\theta\sin\phi & \cos\theta \end{pmatrix}$$

In this case we can have $0 \le \phi < 2\pi$ and also $0 \le \psi < 2\pi$, but to avoid duplication we restrict $0 \le \theta < \pi$. However we cannot avoid duplication entirely since when $\theta = 0$ we get the same matrix whenever $\phi + \psi$ has a constant value. This is easily seen from the diagram defining the Euler angles. We will see later that this case is important when studying robot wrists.

2.6 General Rigid Motions in 3-D

Pure translations can be dealt with very quickly, since they are so similar to the 2-D case. Again translations can be represented by vectors, three component vectors this time of course. Again rotations act on translations and we may write a general 3-D rigid motion as a 4×4 partitioned matrix:-

$$\left(\begin{array}{c|c} \mathbf{R}(\phi, \hat{\mathbf{v}}) & \mathbf{t} \\ \hline 0 & 1 \end{array} \right)$$

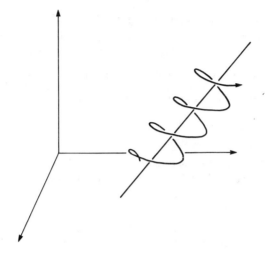

Figure 2.6 A General Screw Motion

This again models the 'funny' multiplication given by the action of the rotations on the translations:-

$$\left(\begin{array}{c|c}\mathbf{R}(\phi_2,\hat{\mathbf{v}}_2) & \mathbf{t}_2 \\ \hline 0 & 1\end{array}\right)\left(\begin{array}{c|c}\mathbf{R}(\phi_1,\hat{\mathbf{v}}_1) & \mathbf{t}_1 \\ \hline 0 & 1\end{array}\right) =$$
$$\left(\begin{array}{c|c}\mathbf{R}(\phi_2,\hat{\mathbf{v}}_2)\mathbf{R}(\phi_1,\hat{\mathbf{v}}_1) & \mathbf{R}(\phi_2,\hat{\mathbf{v}}_2)\mathbf{t}_1 + \mathbf{t}_2 \\ \hline 0 & 1\end{array}\right)$$

We saw earlier that a general 2-D rigid motion is (almost) always a rotation about some point in the plane. For 3-D rigid motions we might be tempted to think that the general motion would be a rotation about some line in space. This, however, turns out not to be general enough: we must also allow a translation along the rotation axis. The result of such a rotation plus translation is a helical or screw motion, see fig. 2.6.

A screw motion is given by the following matrix:-

$$\left(\begin{array}{c|c}\mathbf{I} & \mathbf{u} \\ \hline 0 & 1\end{array}\right)\left(\begin{array}{c|c}\mathbf{R}(\phi,\hat{\mathbf{v}}) & \frac{\phi p}{2\pi}\hat{\mathbf{v}} \\ \hline 0 & 1\end{array}\right)\left(\begin{array}{c|c}\mathbf{I} & -\mathbf{u} \\ \hline 0 & 1\end{array}\right) = \left(\begin{array}{c|c}\mathbf{R}(\phi,\hat{\mathbf{v}}) & \frac{\phi p}{2\pi}\hat{\mathbf{v}} + (\mathbf{I} - \mathbf{R}(\phi,\hat{\mathbf{v}}))\mathbf{u} \\ \hline 0 & 1\end{array}\right)$$

Here, the middle matrix is a screw about a line through the origin; that is, a rotation with axis $\hat{\mathbf{v}}$ followed by a translation along $\hat{\mathbf{v}}$. The outer matrices conjugate the screw and serve to place the line at an arbitrary position in space. The parameter p is the **pitch** of the screw, it gives the distance advanced along the axis for every complete turn, exactly like the pitch on the thread of an ordinary nut or bolt. When the pitch is zero the screw is a pure rotation, positive pitches correspond to left-hand threads and negative pitch to right-handed threads.

To show that a general rigid motion is a screw motion, we must show how to put a general transformation into the form derived above. This amounts to being able to find the parameters $\hat{\mathbf{v}}$, $\mathbf{u}$ and p. The unit vector in the direction of the line $\hat{\mathbf{v}}$ is easy since it must be the eigenvector of the rotation matrix corresponding to the unit eigenvalue. (This fails if $\mathbf{R} = \mathbf{I}$, that is if the motion is a pure translation.) The vector $\mathbf{u}$ is more difficult to find since it is the position vector of any point on the rotation axis. However we can uniquely specify $\mathbf{u}$ by requiring that it is normal to the rotation axis. So we impose the extra restriction that $\hat{\mathbf{v}} \cdot \mathbf{u} = 0$. So to put the general matrix $\left(\begin{array}{c|c} \mathbf{R} & \mathbf{t} \\ \hline 0 & 1 \end{array} \right)$ into the above form we must solve the following system of linear equations:-

$$\frac{\phi p}{2\pi}\hat{\mathbf{v}} + (\mathbf{I} - \mathbf{R})\mathbf{u} = \mathbf{t}$$

Now $\hat{\mathbf{v}} \cdot \mathbf{R}\mathbf{u} = \hat{\mathbf{v}} \cdot \mathbf{u} = 0$, since the rotation is about $\hat{\mathbf{v}}$. So we can dot the above equation with $\hat{\mathbf{v}}$ to give $0 = \hat{\mathbf{v}} \cdot (\mathbf{t} - \frac{\phi p}{2\pi}\hat{\mathbf{v}})$. This enables us to find the pitch:-

$$p = \frac{2\pi}{\phi}\hat{\mathbf{v}} \cdot \mathbf{t}$$

All we have to do now is to solve:-

$$(\mathbf{I} - \mathbf{R})\mathbf{u} = (\mathbf{t} - (\hat{\mathbf{v}} \cdot \mathbf{t})\hat{\mathbf{v}})$$

This is possible even though $\det(\mathbf{I} - \mathbf{R}) = 0$, since the equations will be consistent.

As an example we will find the axis and pitch of the following rigid transformation:-

$$\begin{pmatrix} \frac{2+\sqrt{3}}{4} & \frac{2-\sqrt{3}}{4} & \frac{1}{2\sqrt{2}} & \frac{-1}{6\sqrt{2}} \\ \frac{2-\sqrt{3}}{4} & \frac{2+\sqrt{3}}{4} & \frac{-1}{2\sqrt{2}} & \frac{5}{6\sqrt{2}} \\ \frac{-1}{2\sqrt{2}} & \frac{1}{2\sqrt{2}} & \frac{\sqrt{3}}{2} & \frac{2-\sqrt{3}}{2} \\ 0 & 0 & 0 & 1 \end{pmatrix}$$

First we must find $\hat{\mathbf{v}} = (v_x, v_y, v_z)^T$ and ϕ, we can do this by solving the equation $(\mathbf{I} - \mathbf{R})\hat{\mathbf{v}} = 0$, since we know that $\hat{\mathbf{v}}$, the axis of rotation, is unchanged by the matrix:-

$$\begin{pmatrix} \frac{2-\sqrt{3}}{4} & \frac{-2+\sqrt{3}}{4} & \frac{-1}{2\sqrt{2}} \\ \frac{-2+\sqrt{3}}{4} & \frac{2-\sqrt{3}}{4} & \frac{1}{2\sqrt{2}} \\ \frac{1}{2\sqrt{2}} & \frac{-1}{2\sqrt{2}} & \frac{2-\sqrt{3}}{2} \end{pmatrix} \begin{pmatrix} v_x \\ v_y \\ v_z \end{pmatrix} = \begin{pmatrix} 0 \\ 0 \\ 0 \end{pmatrix}$$

This homogeneous system of equations has a solution since the matrix on the left-hand side is singular; this is easily seen, as the first row is the negative of the second. If we multiply the second row by $\frac{1}{2\sqrt{2}}$ and subtract $\frac{2+\sqrt{3}}{4}$ times the third row we get $(\frac{-3}{4} + \frac{\sqrt{3}}{2})v_z = 0$. Hence, $v_z = 0$; substituting this in any of the equations gives $v_x = v_y$. So the unit vector is:-

$$\hat{\mathbf{v}} = \begin{pmatrix} \frac{1}{\sqrt{2}} \\ \frac{1}{\sqrt{2}} \\ 0 \end{pmatrix}$$

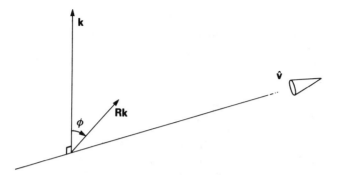

Figure 2.7 Finding the Angle of Rotation

To find the angle of rotation ϕ, observe that $\hat{\mathbf{v}}$ has no component in the z-direction, so $\mathbf{k} \cdot \hat{\mathbf{v}} = 0$. Since $\mathbf{k}$ is perpendicular to the rotation axis, the effect of the rotation is simply to rotate it about the axis, see fig. 2.7. Therefore, $\mathbf{k} \cdot \mathbf{R}\,\mathbf{k} = \cos\phi$.

So $\cos\phi = \sqrt{3}/2$ and hence $\phi = \pi/6$. We should really check that the angle is $\pi/6$, not $-\pi/6$. This would be equivalent to reversing the sign of $\hat{\mathbf{v}}$. A quick way to check this is by looking at the scalar triple product $\hat{\mathbf{v}} \cdot \mathbf{R}\,\mathbf{k} \wedge \mathbf{k}$: if this is positive our results are correct. This works because, by definition, the vector product $\mathbf{a} \wedge \mathbf{b}$ gives a vector perpendicular to both $\mathbf{a}$ and $\mathbf{b}$ but where if one looks along the vector then $\mathbf{a}$ must be turned clockwise to bring it into coincidence with $\mathbf{b}$. Our rotations are positive when anticlockwise, hence the ordering in the product; $\mathbf{R}\,\mathbf{k} \wedge \mathbf{k}$.

The pitch is simply given by:-

$$p = \frac{2\pi}{\phi}(\hat{\mathbf{v}} \cdot \mathbf{t}) = 4$$

measured as length per radian. Finally, $\mathbf{u}$ satisfies $(\mathbf{I} - \mathbf{R})\mathbf{u} = \mathbf{t} - \frac{\phi p}{2\pi}\hat{\mathbf{v}}$.

$$\begin{pmatrix} \frac{2-\sqrt{3}}{4} & \frac{-2+\sqrt{3}}{4} & \frac{-1}{2\sqrt{2}} \\ \frac{-2+\sqrt{3}}{4} & \frac{2-\sqrt{3}}{4} & \frac{1}{2\sqrt{2}} \\ \frac{1}{2\sqrt{2}} & \frac{-1}{2\sqrt{2}} & \frac{2-\sqrt{3}}{2} \end{pmatrix} \begin{pmatrix} u_x \\ u_y \\ u_z \end{pmatrix} = \begin{pmatrix} \frac{-1}{2\sqrt{2}} \\ \frac{1}{2\sqrt{2}} \\ \frac{2-\sqrt{3}}{2} \end{pmatrix}$$

The solution to this singular system can be found in the same way as the solution to the homogeneous system above. The general solution is:-

$$\mathbf{u} = \begin{pmatrix} 0 \\ 0 \\ 1 \end{pmatrix} + \lambda \begin{pmatrix} \frac{1}{\sqrt{2}} \\ \frac{1}{\sqrt{2}} \\ 0 \end{pmatrix}$$

The solution perpendicular to the axis is therefore $\mathbf{u} = (0, 0, 1)^T$.

In this chapter we have seen how to write a general rigid motion in 3-D as a 4×4 matrix. In general such a motion is a screw motion, and we showed this by demonstrating how to find the axis and pitch of the screw given a four-matrix.

Exercises

2.4 A rigid body is rotated $\frac{\pi}{2}$ radians about the x-axis and then $\frac{\pi}{2}$ radians about the y-axis. Find the axis of the resulting composite rotation. Also, find the axis of the result when the rotation about the y-axis is performed first.

2.5 Find the 4×4 matrices corresponding to the following 3-D rigid transformations:

(i)   A rotation of $\frac{\pi}{4}$ radians about the x-axis, followed by a translation of 2 units in the z-direction.

(ii) A translation of 2 units in the z-direction, followed by a rotation of $\frac{\pi}{4}$ radians about the x-axis.

(iii) A translation of 2 units in the z-direction, followed by a rotation of $\frac{\pi}{4}$ radians about the x-axis, followed by a translation of -2 units in the z-direction.

2.6 A rigid body transformation takes the three points:-

$$(0,0,0), \quad (1,0,0) \quad \text{and} \quad (0,1,0)$$

to the points:-

$$(1,-1,1), \quad (1,0,1) \quad \text{and} \quad (0,-1,1)$$

respectively. Find the 4×4 matrix corresponding to this transformation. Also find the axis and pitch of the motion.

2.7 Let $\mathbf{R}$ be a 3×3 rotation matrix and let $\mathbf{A} = \mathbf{R} - \mathbf{R}^T$:

(i)   Show that $\mathbf{A}$ is an antisymmetric matrix (that is, a matrix which satisfies $\mathbf{A}^T = -\mathbf{A}$, also called a skew symmetric matrix).

(ii) Show that $\mathbf{A} = 0$ if and only if $\mathbf{R}$ is a rotation of 0 or π radians.

(iii) Show that if $\hat{\mathbf{v}}$ is an eigenvector of $\mathbf{R}$ with non-zero eigenvalue γ then $\hat{\mathbf{v}}$ is also an eigenvector of $\mathbf{R}^T$ but with eigenvalue $\frac{1}{\gamma}$.

(iv) Hence, show that the matrix $\mathbf{A}$ given in (i) above is of the form:-

$$\mathbf{A} = \lambda \begin{pmatrix} 0 & v_z & -v_y \\ -v_z & 0 & v_x \\ v_y & -v_x & 0 \end{pmatrix}$$

Where λ is an arbitrary constant and v_x, v_y and v_z are the components of the eigenvector $\hat{\mathbf{v}}$, corresponding to the eigenvalue 1.

Note, this gives a very easy way to find the axis of a 3×3 rotation matrix as long as the angle of rotation θ is not 0 or π radians. In fact the constant can be shown to be $\lambda = 2\sin\theta$.

3 Robot Anatomy

In this section we will look at the basic parts which make up robots. Essentially we consider robots to be made up of rigid links connected together with joints. In the simplest cases the links are connected together in series, to form an open loop structure. However, more complicated arrangements are sometimes used, and the analysis of manipulators containing closed loops is more complex. First we look at the links and see how to describe their positions and orientations.

3.1 Links

Since we are considering the links to be rigid bodies, we saw in the previous chapter how to specify their transformations. As already mentioned, we may define a 'home' configuration for the link. Then subsequent positions are described by giving the rigid body transformation which moves the link from 'home' to the new configuration. This involves specifying six parameters, three Euler angles (or equivalent) to determine the orientation of the link, and three components of a translation vector to give the position. Hence we say that an unconstrained rigid body has six degrees-of-freedom. This means that changes in any of the six parameters will result in a change of configuration of the link which is independent of changes in the other parameters.

Our purpose here is the other way around, that is, to find the rigid transformation which takes the link from one given configuration to another. The shape and mass distribution of the links are unimportant here. We can find all we need to know about a link by observing three points on it, see fig. 3.1; for the 2-D case only two points are needed, see exercise 2.3. Suppose the three points have position vectors p_1, p_2 and p_3 in the home configuration. Then after a rigid body transformation the new position vectors of the points will be:-

$$\mathbf{p}_1' = \mathbf{R}\,\mathbf{p}_1 + \mathbf{t}, \quad \mathbf{p}_2' = \mathbf{R}\,\mathbf{p}_2 + \mathbf{t}, \quad \mathbf{p}_3' = \mathbf{R}\,\mathbf{p}_3 + \mathbf{t} \tag{$*$}$$

Given the initial and final positions we can work out what rigid transformation must

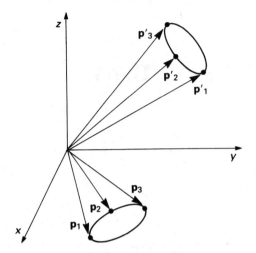

Figure 3.1 Rigid Body Motion Specified by Three Points

have been performed. For example suppose:-

$$\mathbf{p}_1 = \begin{pmatrix} 0 \\ 0 \\ 0 \end{pmatrix}, \quad \mathbf{p}_2 = \begin{pmatrix} 1 \\ 0 \\ 0 \end{pmatrix}, \quad \mathbf{p}_3 = \begin{pmatrix} 0 \\ 1 \\ 0 \end{pmatrix}$$

and

$$\mathbf{p}'_1 = \begin{pmatrix} 1 \\ -1 \\ 1 \end{pmatrix}, \quad \mathbf{p}'_2 = \begin{pmatrix} 1 \\ 0 \\ 1 \end{pmatrix}, \quad \mathbf{p}'_3 = \begin{pmatrix} 0 \\ -1 \\ 1 \end{pmatrix}$$

Now we must solve the equations (∗) above for $\mathbf{R}$ and $\mathbf{t}$. In this case, it is quite easy to solve for $\mathbf{t}$, since $\mathbf{R}\,\mathbf{p}_1 = 0$ as $\mathbf{p}_1$ is already the zero vector. Hence, $\mathbf{t} = \mathbf{p}'_1$. So we are left with two equations:-

$$\mathbf{R}\,\mathbf{p}_2 = \mathbf{p}'_2 - \mathbf{p}'_1, \quad \mathbf{R}\,\mathbf{p}_3 = \mathbf{p}'_3 - \mathbf{p}'_1$$

Again since $\mathbf{p}_2$ and $\mathbf{p}_3$ have a particularly simple form it is easy to find the first two columns of $\mathbf{R}$. Let:-

$$\mathbf{R} = \begin{pmatrix} r_{11} & r_{12} & r_{13} \\ r_{21} & r_{22} & r_{23} \\ r_{31} & r_{32} & r_{33} \end{pmatrix}$$

then:-

$$\begin{pmatrix} r_{11} \\ r_{21} \\ r_{31} \end{pmatrix} = \mathbf{p}'_2 - \mathbf{p}'_1, \quad \begin{pmatrix} r_{12} \\ r_{22} \\ r_{32} \end{pmatrix} = \mathbf{p}'_3 - \mathbf{p}'_1$$

To find the third column we need another equation. This can be done by taking the vector product of the two we have just used:-

$$\mathbf{R}\,\mathbf{p}_2 \wedge \mathbf{R}\,\mathbf{p}_3 = \mathbf{R}\,(\mathbf{p}_2 \wedge \mathbf{p}_3) = (\mathbf{p}'_2 - \mathbf{p}'_1) \wedge (\mathbf{p}'_3 - \mathbf{p}'_1)$$

Now, in this case, $(\mathbf{p}_2 \wedge \mathbf{p}_3) = (0, 0, 1)^T$, and so:-

$$\begin{pmatrix} r_{13} \\ r_{23} \\ r_{33} \end{pmatrix} = (\mathbf{p}_2' - \mathbf{p}_1') \wedge (\mathbf{p}_3' - \mathbf{p}_1')$$

Notice that this also ensures that the rotation matrix satisfies the relation $\mathbf{R}^T\mathbf{R} = \mathbf{I}$. That is, the columns of the matrix are mutually orthogonal unit vectors.

Putting all this together we have:-

$$\mathbf{R} = \begin{pmatrix} 0 & -1 & 0 \\ 1 & 0 & 0 \\ 0 & 0 & 1 \end{pmatrix}, \qquad \mathbf{t} = \begin{pmatrix} 1 \\ -1 \\ 1 \end{pmatrix}$$

The answer is a $\frac{\pi}{2}$ rotation about the z-axis together with a translation of one unit in each of the x, $-y$ and z directions.

Choosing 'nice' points simplifies the calculations here. It is possible to find the rigid transformation from any three points, as long as they do not all lie on a line. Essentially we must solve a system of linear equations for the twelve unknowns; the r_{ij}'s and t_i's. These equations will be linearly independent so long as the three points are not colinear. Hence we can keep track of links by following the progress of three points fixed to the link.

3.2 Joints

Consider how we can join links together. At first sight there seem to be limitless ways of attaching one link to another while still allowing relative movement. In the 1870s, Franz Reuleaux, a German mechanical engineer, simplified things by defining lower pairs, see fig. 3.2. A **Reuleaux lower pair** is a pair of identical surfaces; one solid, the other hollow. These surfaces fit together but can still move relative to each other while remaining in contact.

Reuleaux found six such pairs, and it can be shown that these are the only possibilities.

- Any surface of revolution gives a revolute or R-pair.

- Any helicoidal surface, like the mating surfaces of a nut and bolt, give a screw or H-pair.

- Any surface of translation, like a prism, results in a prismatic or P-pair.

- The surface of a cylinder is a surface of rotation and translation. Two cylinders form a cylindric or C-pair.

- A sphere is a surface of revolution about any diameter. A ball and socket are a spherical or S-pair.

- A plane is a surface of translation about any line in the plane and also a surface of revolution about any normal line. Two planes form a planar or E-pair.

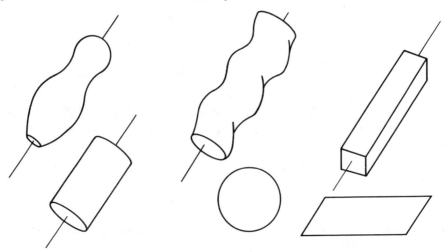

Figure 3.2 The Reuleaux Lower Pairs

Each of these pairs can be used as a joint. Just fix one of the mating surfaces to one link and the other to the second link. For example, the revolute pair will give a simple hinge joint between two links. In fact any kind of articulation between links can be thought of as combinations of these six, at least infinitesimally. This is because, as we have seen, any rigid motion can be thought of as a screw motion or a translation. We will regard these as our fundamental joints and will not consider any others. In fact we could do everything in terms of just helicoidal and prismatic joints. Each H-joint has a pitch associated with it, the pitch of the screw thread. So the revolute joint is just a pitch zero H-joint.

These joints are one degree-of-freedom joints, that is, we need one parameter to give the relative position of the two sides of the joint. Such parameters are called joint variables. The cylindric joint is a two degree-of-freedom joint, since it allows both rotations about, and translations along, an axis, thus needing two parameters. Planar and spherical joints are three degree-of-freedom joints. The spherical joint allows rotations about a point, and we need three parameters to describe such rotations, the Euler angles for example, see section 2.5. The planar joint will allow movements only in a plane. To specify such moves we need three parameters: the angle of rotation about a fixed point and the lengths of translation along two orthogonal directions will do.

Given a joint, what rigid motions can the links execute with respect to each other? For generality let us assume we have a screw joint. Hold one link fixed, with the axis of the screw aligned along the x-axis for convenience. See fig. 3.3. The second link can undergo a screw motion about the axis given by:-

$$\mathbf{A}\left(\lambda\right) = \begin{pmatrix} 1 & 0 & 0 & \frac{\lambda p}{2\pi} \\ 0 & \cos\lambda & -\sin\lambda & 0 \\ 0 & \sin\lambda & \cos\lambda & 0 \\ 0 & 0 & 0 & 1 \end{pmatrix}$$

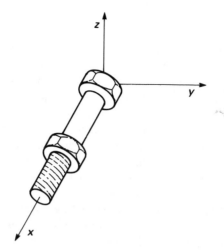

Figure 3.3 Rigid Motions Allowed by a Screw Joint

where p is the pitch of the screw.

The possible rigid motions that can be performed by a screw joint form a one parameter family. The single parameter is the joint variable λ. If the joint is a revolute joint then $p = 0$ and thus the possible rigid motions are given by:-

$$\mathbf{A}(\lambda) = \begin{pmatrix} 1 & 0 & 0 & 0 \\ 0 & \cos\lambda & -\sin\lambda & 0 \\ 0 & \sin\lambda & \cos\lambda & 0 \\ 0 & 0 & 0 & 1 \end{pmatrix}$$

The joint variable here is an angle. For a prismatic joint aligned along the x-axis the corresponding rigid transformations are:-

$$\mathbf{A}(\lambda) = \begin{pmatrix} 1 & 0 & 0 & \lambda \\ 0 & 1 & 0 & 0 \\ 0 & 0 & 1 & 0 \\ 0 & 0 & 0 & 1 \end{pmatrix}$$

Here the joint variable is a length.

If the joint is positioned arbitrarily in space, then we can find the rigid transformations allowed by conjugating the results above, see also section 2.6. Suppose $\mathbf{T} = \left(\begin{array}{c|c} \mathbf{R} & \mathbf{t} \\ \hline 0 & 1 \end{array} \right)$ is a rigid transformation which takes the x-axis to the joint axis. The new one parameter family of motions will be given by:-

$$\mathbf{A}'(\lambda) = \mathbf{T}\mathbf{A}(\lambda)\mathbf{T}^{-1}$$

3.3   Geometric Design

If a robot is to move objects around generally, it will need six degrees-of-freedom, since it should be able to position and orient the object. This can be achieved by connecting six one degree-of-freedom joints in series. The position and orientation of the final link, which carries the object, will be specified by six parameters, the six joint variables $(\theta_1, \theta_2, \ldots, \theta_6)$.

The joint variables can be thought of as co-ordinates of a space; the **joint space** of the robot. Each different configuration of the robot corresponds to a different point in joint space. Furthermore, every point in joint space corresponds to a configuration of the robot. In a practical robot the range of movement of each joint will be limited. For example real revolute joints cannot usually turn a full circle but can only rotate in a limited range, say $-2.3 \geq \theta \geq +2.3$ radians. Hence the practical robot will only be able to achieve configurations corresponding to a subspace of the total joint space.

The **work space** of the robot, by contrast, is the space of positions and orientations reachable by the robot's end-effector. This will be a subspace of all possible rigid body transformations, since there will always be transformations which the robot cannot perform. For instance a robot of finite size will not be able to perform translations over an arbitrarily long distance. Note, however, that some writers consider a robot's work space to be only the space of positions reachable by some point on the end-effector, irrespective of the possible orientations. If this is done it is necessary to specify which particular point on the effector one is considering, by giving its home coordinates for example.

Some robots have only five degrees-of-freedom. In applications such as paint spraying and arc welding, the final orientation about the axis of the tool is irrelevant. Assembly robots often only require four degrees-of-freedom. They need three degrees-of-freedom to position and orient objects in the plane and an extra degree-of-freedom to lift objects out of the plane.

Although many different designs for the arrangement of the joints in a robot are possible, only a few have ever been used. This is because the inverse kinematics of the more esoteric design cannot be solved; see later. The most common designs are shown in fig. 3.4. All of the current designs separate the positioning from the orientation. The orientation is taken care of by three joints in the form of a 'wrist', while another three joints are able to position the centre of the wrist. This does make the analysis far easier but is not the most general design possible. Many of the designs are constructed so that each of the first three joints controls the position of the last link along one co-ordinate axis; not necessarily in cartesian coordinates though. Since electric motors are readily available and easily controlled components, revolute joints are often preferred to prismatic ones, and the 'Puma' type design is very common. 'Puma' is an acronym for Programmable Universal Machine for Assembly. Its design is based on the human arm and hence is sometimes referred to as the anthropomorphic design. In special circumstances prismatic joints are used, for example in large gantry type robots.

So far all the designs considered are open loop designs. Closed loops are very important in robotics. Not only are there robot designs which incorporate closed loops but also if an open loop robot follows a definite path, then it can be considered as a closed loop. For example, consider a Puma type robot performing a welding operation on a large circular

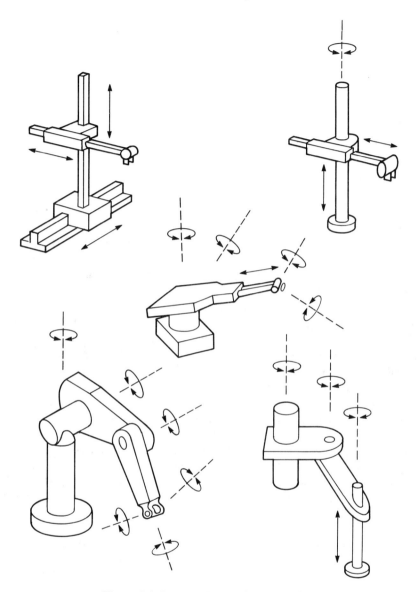

Figure 3.4 Common Types of Robot Design

pipe; see fig. 3.5.

Because three of the robot's axes are parallel and also parallel to the axis of the pipe, the robot moves like a four-bar mechanism. In mechanical engineering a mechanism is a movable system of links and joints, which can contain several closed loops. Mechanisms usually, but not always, have only one degree-of-freedom, that is, the movement of any of

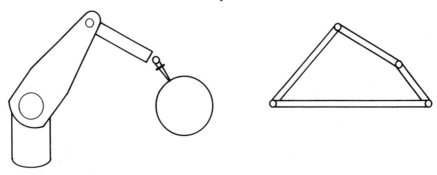

Figure 3.5 Puma Robot Welding a Pipe, and a Four-bar

the links or bars can be parameterized by one variable. The four-bar mechanism is probably the most common mechanism used in practice; its use and study have a long history. As far as mechanics is concerned, mechanisms and robots are identical, so much of the material of this book also applies to mechanisms. Similarly, work on mechanisms can be useful in robotics.

The most common robot employing a closed loop is the parallelogram type, see fig. 3.6. In fact, this is almost identical to the Puma from the point of view of kinematics. However, this design allows the motor for the third joint to be mounted near the base, not on the second link. Thus the end-effector can carry a heavier payload. For this reason other closed loop designs are the subject of current research, with the hope that more of the drive motors can be moved to the base of the robot.

Finally, the Stewart platform, shown in fig. 3.6, is not usually considered to be a robot, in which case we must call it a six degrees-of-freedom mechanism. Its main use is in aircraft simulators. The pilot sits in the simulated cockpit on the platform, while the six hydraulic rams move the platform to mimic the plane's motion. Each ram is connected to the base and the platform by a passive (not driven) spherical joint. This allows movement in all but the direction parallel to the ram, hence each ram imparts one degree-of-freedom to the platform. Because none of the rams are in series, this mechanism is often referred to as a

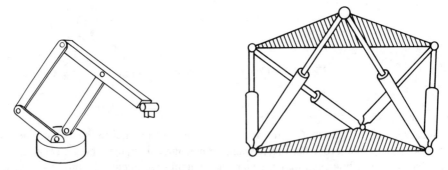

Figure 3.6 Parallelogram Design and the Stewart Platform

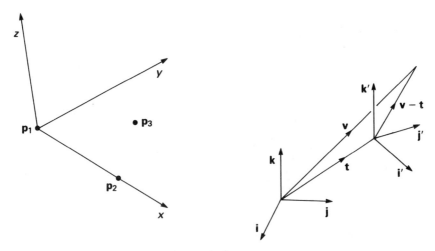

Figure 3.7 Coordinate Frames

parallel manipulator.

3.4 Co-ordinate Frames

In robotics it is common to keep track of link positions using embedded co-ordinate frames. This method and the three point methods described above are equivalent. This is because we can always use three points to define a co-ordinate frame. We simply take $\mathbf{p}'_1$ as the origin of the new frame, fix the new x-axis along the direction of the vector $\mathbf{p}'_2 - \mathbf{p}'_1$, and finally choose the new xy-plane so that it contains the point with position vector $\mathbf{p}'_3$; see fig. 3.7.

The problem with using embedded co-ordinate frames is that it is often very confusing to deal with several different co-ordinate frames. Also one must introduce a new kind of transformation. So far we have only had one co-ordinate frame and we have studied the transformations we need to apply to vectors to move them around. This is called the active point of view and the transformations are called active transformations. The other point of view is the passive point of view. Here the vector is fixed and the co-ordinate frame moves. A passive transformation will give us the components of the vector in the new co-ordinates.

Let us assume that $\mathbf{i}, \mathbf{j}$ and $\mathbf{k}$ are respectively the unit vectors in the x, y and z directions. Now we may change co-ordinates so that the new basis vectors are given by:-

$$\mathbf{i}' = \mathbf{R}\,\mathbf{i}$$
$$\mathbf{j}' = \mathbf{R}\,\mathbf{j}$$
$$\mathbf{k}' = \mathbf{R}\,\mathbf{k}$$

Further, suppose the new origin has position vector $\mathbf{t}$, with respect to the old origin.

Consider a point with position vector $\mathbf{v} = (x, y, z)^T$, relative to the old origin, see fig. 3.7. With respect to the new origin the point has position vector $\mathbf{v} - \mathbf{t}$. In the new co-ordinates this is:-

$$
\begin{array}{rcccl}
x' &=& \mathbf{i}' \cdot (\mathbf{v} - \mathbf{t}) &=& \mathbf{i} \cdot (\mathbf{R}^T\mathbf{v} - \mathbf{R}^T\mathbf{t}) \\
y' &=& \mathbf{j}' \cdot (\mathbf{v} - \mathbf{t}) &=& \mathbf{j} \cdot (\mathbf{R}^T\mathbf{v} - \mathbf{R}^T\mathbf{t}) \\
z' &=& \mathbf{k}' \cdot (\mathbf{v} - \mathbf{t}) &=& \mathbf{k} \cdot (\mathbf{R}^T\mathbf{v} - \mathbf{R}^T\mathbf{t})
\end{array}
$$

In terms of the old co-ordinates this can be written as:-

$$
\begin{pmatrix} x' \\ y' \\ z' \\ 1 \end{pmatrix} = \left(\begin{array}{c|c} \mathbf{R}^T & -\mathbf{R}^T\mathbf{t} \\ \hline 0 & 1 \end{array} \right) \begin{pmatrix} x \\ y \\ z \\ 1 \end{pmatrix}
$$

Notice that the 4×4 matrix which effects the transformation is the inverse of the active transformations we have already considered.

We have seen two methods for keeping track of links and other rigid bodies. In the end it is a matter of taste whether one uses the active or passive viewpoint. The active viewpoint is preferred in this book. We have also seen how to write down the transformations allowed by joint, as a matrix with a variable joint parameter. Finally, we have introduced the concepts of joint space and work space for open loop robots.

Exercises

3.1 Find the rigid body transformations which take the points $(0, 0, 0)$, $(1, 0, 0)$ and $(0, 1, 0)$ respectively to:-

(i) $(2, 0, 0)$, $(3, 0, 0)$ and $(2, 0, 1)$

(ii) $(0, 0, 0)$, $(0, 1, 0)$ and $(-1, 0, 0)$

(iii) $(0, 0, 1)$, $(0, 1, 1)$ and $(-1, 0, 1)$

Find also the pitch and axis of these transformations.

3.2 Find the possible rigid body motions that can be generated by the following joints:-

(i)   A revolute joint aligned with the z-axis.

(ii) A prismatic joint aligned with the z-axis.

(iii) A revolute joint located along the line $x = 1$, $y = 0$.

(iv) A screw joint of pitch 2 aligned with $x = 1$, $y = 0$.

3.3 Let $\mathbf{A} = \left(\begin{array}{c|c} \mathbf{R} & \mathbf{v} \\ \hline 0 & 1 \end{array} \right)$ be a rigid body transformation; verify that the inverse transformation is given by:-

$$
\mathbf{A}^{-1} = \left(\begin{array}{c|c} \mathbf{R}^T & -\mathbf{R}^T\mathbf{v} \\ \hline 0 & 1 \end{array} \right)
$$

3.4   Consider a rigid transformation $\left(\dfrac{\mathbf{R}\mid\mathbf{t}}{0\mid 1}\right)$, where $\mathbf{R}$ is a rotation of θ radians about the vector $\hat{\mathbf{v}}$. Let $\mathbf{u}$ be the position vector of a point on the axis of this transformation. Show that:-

$$(\mathbf{I} + \mathbf{R}^T)\mathbf{t} = 2\sin\theta\,\mathbf{u} \wedge \hat{\mathbf{v}}$$

Now, suppose $\hat{\mathbf{v}}_a$ and $\hat{\mathbf{v}}_b$ are two orthogonal unit vectors perpendicular to $\hat{\mathbf{v}}$ and they satisfy $\hat{\mathbf{v}}_a \wedge \hat{\mathbf{v}}_b = \hat{\mathbf{v}}$. In other words assume $\hat{\mathbf{v}}_a$, $\hat{\mathbf{v}}_b$, $\hat{\mathbf{v}}$ form a set of orthonormal basis vectors for some right-handed co-ordinate frame. Show that the projections of $\mathbf{u}$ on to these two vectors is given by:-

$$\begin{aligned}
\mathbf{u}\cdot\hat{\mathbf{v}}_a &= \frac{1}{2\sin\theta}\hat{\mathbf{v}}_b^T(\mathbf{I}+\mathbf{R}^T)\mathbf{t} \\
\mathbf{u}\cdot\hat{\mathbf{v}}_b &= \frac{-1}{2\sin\theta}\hat{\mathbf{v}}_a^T(\mathbf{I}+\mathbf{R}^T)\mathbf{t}
\end{aligned}$$

4   Kinematics

Kinematics is the study of possible movement and configurations of a system. It is really only concerned with the geometry of the system. To understand how the system **will** move in a given circumstance requires a knowledge of forces, inertias, energy and so forth. This is the subject matter of dynamics which will be discussed later.

For robots we need to know the position and orientation of the last link or end-effector, in terms of the joint variables. This is the **forward kinematics**. Before we look at a six axis industrial robot we will look at some smaller examples; the three joint planar manipulator and the three joint spherical manipulator. These are not, however, trivial examples. Many commercial robots incorporate these structures. Any three parallel R-joints form a planar manipulator; see, for example, the Puma and Scara type robots. Also any three R-joints with intersecting axes will form a spherical manipulator; such structures are used as wrists in many robots.

The basic idea is to find the matrices corresponding to the motions about each joint. We met these matrices in section 3.2. Combining these matrices in the correct order will give the rigid body transformations of the final link as a function of the joint variables.

4.1   The Planar Manipulator

In the plane three joints are needed to give the manipulator three degrees-of-freedom. To specify the position of a link in the plane we could use two points, but we can also use one point (x, y) and an angle Φ. The joint variables will be three angles θ_1, θ_2 and θ_3. As usual, positive angles denote anticlockwise rotations. We must also specify the lengths of the links, we will leave these 'design parameters' arbitrary and just write l_1, l_2 and l_3; see fig. 4.1.

The first thing to do is to define a suitable home configuration. In this case a convenient position would be where all the joint angles are zero, $\theta_1 = \theta_2 = \theta_3 = 0$. See fig. 4.1. Now to get to the configuration with joint angles $(\theta_1, \theta_2, \theta_3)$, we perform the following three rotations.

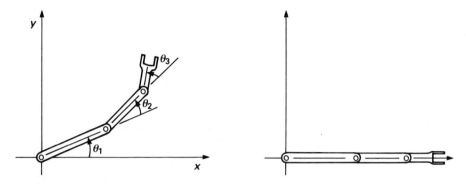

Figure 4.1 The Planar Manipulator and its Home Configuration

- First rotate joint 3 to angle θ_3. This is given by a 3×3 matrix of the form:-

$$\mathbf{A}_3(\theta_3) = \left(\frac{\mathbf{R}(\theta_3) \mid \mathbf{t}_3}{0 \mid 1} \right)$$

The centre of rotation is the position of the third joint in the home position and has co-ordinates $(l_1 + l_2, 0)$. So the matrix is:-

$$\mathbf{A}_3(\theta_3) = \begin{pmatrix} \cos\theta_3 & -\sin\theta_3 & (1-\cos\theta_3)(l_1+l_2) \\ \sin\theta_3 & \cos\theta_3 & -\sin\theta_3(l_1+l_2) \\ 0 & 0 & 1 \end{pmatrix}$$

Notice that this rotation does not change the positions of the first and second joints. So the second stage is:-

- Rotate joint two, to angle θ_2. The matrix for this is just:-

$$\mathbf{A}_2(\theta_2) = \begin{pmatrix} \cos\theta_2 & -\sin\theta_2 & (1-\cos\theta_2)l_1 \\ \sin\theta_2 & \cos\theta_2 & -\sin\theta_2 l_1 \\ 0 & 0 & 1 \end{pmatrix}$$

since the centre of rotation is at $(l_1, 0)$.

- Finally, rotate the first joint to angle θ_1. Since this axis has not been affected by the previous rotations the matrix is simply:-

$$\mathbf{A}_1(\theta_1) = \begin{pmatrix} \cos\theta_1 & -\sin\theta_1 & 0 \\ \sin\theta_1 & \cos\theta_1 & 0 \\ 0 & 0 & 1 \end{pmatrix}$$

So the effect of such a movement will be given by the product of the three matrices; remembering that the first operation is the rightmost:-

$$\mathbf{A}_1(\theta_1)\mathbf{A}_2(\theta_2)\mathbf{A}_3(\theta_3) = \begin{pmatrix} \cos(\theta_1+\theta_2+\theta_3) & -\sin(\theta_1+\theta_2+\theta_3) & k_x \\ \sin(\theta_1+\theta_2+\theta_3) & \cos(\theta_1+\theta_2+\theta_3) & k_y \\ 0 & 0 & 1 \end{pmatrix}$$

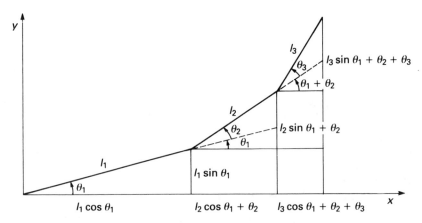

Figure 4.2 Using Trigonometry to Derive the Kinematic Equations

The quantities k_x and k_y are given by:-

$$k_x = l_1 \cos\theta_1 + l_2 \cos(\theta_1 + \theta_2) - (l_1 + l_2)\cos(\theta_1 + \theta_2 + \theta_3)$$
$$k_y = l_1 \sin\theta_1 + l_2 \sin(\theta_1 + \theta_2) - (l_1 + l_2)\sin(\theta_1 + \theta_2 + \theta_3)$$

This matrix $\mathbf{K}(\theta_1, \theta_2, \theta_3) = \mathbf{A}_1(\theta_1)\mathbf{A}_2(\theta_2)\mathbf{A}_3(\theta_3)$ encapsulates everything there is to know about the kinematics of the manipulator. So we will refer to it as the kinematic transformation matrix of the manipulator. The angle Φ is easily seen to be $\Phi = \theta_1 + \theta_2 + \theta_3$. If we want to find the position of any point attached to the end-effector, we simply multiply its position vector in the home position by the matrix $\mathbf{K}$. For example, the point $(l_1 + l_2 + l_3, 0)$ after a general motion will have $x - y$ co-ordinates given by:-

$$\begin{pmatrix} x \\ y \\ 1 \end{pmatrix} = \begin{pmatrix} \cos(\theta_1 + \theta_2 + \theta_3) & -\sin(\theta_1 + \theta_2 + \theta_3) & k_x \\ \sin(\theta_1 + \theta_2 + \theta_3) & \cos(\theta_1 + \theta_2 + \theta_3) & k_y \\ 0 & 0 & 1 \end{pmatrix} \begin{pmatrix} l_1 + l_2 + l_3 \\ 0 \\ 1 \end{pmatrix}$$

This yields the so-called kinematic equations of the manipulator:

$$x = l_1 \cos\theta_1 + l_2 \cos(\theta_1 + \theta_2) + l_3 \cos(\theta_1 + \theta_2 + \theta_3)$$
$$y = l_1 \sin\theta_1 + l_2 \sin(\theta_1 + \theta_2) + l_3 \sin(\theta_1 + \theta_2 + \theta_3)$$

It is a simple matter of trigonometry to check that these equations are correct, see fig. 4.2. However, the matrix approach used above works in all cases, even when the geometry of the situation is not as clear as in the planar case.

4.2   The 3-R Wrist

This structure is used as a wrist in many robots, the Puma and Stanford manipulator, for example. Since all three revolute joint axes meet at a common point, this is an example of

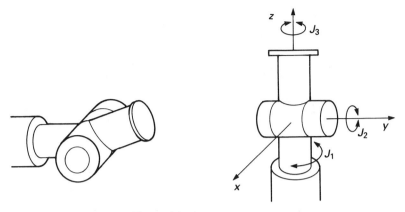

Figure 4.3 The 3-R Wrist

a spherical mechanism. One could think of such a mechanism as being able to manipulate objects on the surface of a sphere. The arrangement of joints (J) is illustrated in fig. 4.3.

For simplicity, the home position has the first and third joints aligned along the z-axis and the second along the y-axis. The origin is chosen to be where the joint axes meet. Hence the kinematic transformation matrix is given by:-

$$\mathbf{K}(\theta_1, \theta_2, \theta_3) = \mathbf{R}(\theta_1, \mathbf{k})\mathbf{R}(\theta_2, \mathbf{j})\mathbf{R}(\theta_3, \mathbf{k})$$

Here we are just using the 3×3 rotation matrices. Notice how similar this is to the Euler angles, and this, of course, is no accident. In full the kinematics are given by:-

$$\mathbf{K} =$$
$$\begin{pmatrix} \cos\theta_1 \cos\theta_2 \cos\theta_3 - \sin\theta_1 \sin\theta_3 & -\cos\theta_1 \cos\theta_2 \sin\theta_3 - \sin\theta_1 \cos\theta_3 & \cos\theta_1 \sin\theta_2 \\ \sin\theta_1 \cos\theta_2 \cos\theta_3 + \cos\theta_1 \sin\theta_3 & -\sin\theta_1 \cos\theta_2 \sin\theta_3 + \cos\theta_1 \cos\theta_3 & \sin\theta_1 \sin\theta_2 \\ -\sin\theta_2 \cos\theta_3 & \sin\theta_2 \sin\theta_3 & \cos\theta_2 \end{pmatrix}$$

Using Euler angles the kinematic equations are just:-

$$\psi = \theta_1, \qquad \theta = \theta_2, \qquad \phi = \theta_3$$

However, if we choose a different parameterization for the rotations, the kinematic equations become extremely complicated and very hard to derive, since we would then have to solve complicated trigonometric equations. The kinematic transform matrix is the simplest way to treat the kinematics of wrists: we can use it to find the positions of points. For example, the point attached to the end-effector and with home position $(0, 0, 1)$ is transformed to:-

$$\mathbf{K} \begin{pmatrix} 0 \\ 0 \\ 1 \end{pmatrix} = \begin{pmatrix} \cos\theta_1 \sin\theta_2 \\ \sin\theta_1 \sin\theta_2 \\ \cos\theta_2 \end{pmatrix}$$

A slight variant of the 3-R wrist is the roll-pitch-yaw wrist, see fig. 4.4. Again this is a spherical mechanism, since all the joint axes intersect. The home configuration might

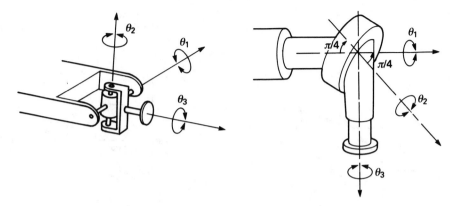

Figure 4.4 The Roll-Pitch-Yaw Wrist and the Cincinnati T^3 Wrist

have the third joint along the x-axis, the second along the y-axis and the first joint along the z-axis. This would lead to the kinematic transform matrix:-

$$\mathbf{K}(\theta_1, \theta_2, \theta_3) = \mathbf{R}(\theta_1, \mathbf{k})\mathbf{R}(\theta_2, \mathbf{j})\mathbf{R}(\theta_3, \mathbf{i})$$

However, if we choose a different home configuration, where the only difference is that $\theta_2' = \theta_2 + \frac{\pi}{2}$, then the wrist is seen to have exactly the same structure as the 3-R wrist above. In practice, the revolute joints will not be fully rotatable, so the home configuration may not be achievable by the manipulator. This does not matter as far as the mathematics is concerned; the choice of home position is a matter of where one chooses the joint variables to be zero.

Exercises

4.1 Consider a three joint planar manipulator with $l_1 = 2, l_2 = 2$ and $l_3 = 1$ in some units. Find the $x - y$ co-ordinates of the point with home position $(5, 0)$, and the angle the last link makes with the x-axis when the joint angles are:-

(i) $\theta_1 = \pi/6,$ $\theta_2 = \pi/6,$ $\theta_3 = \pi/6$

(ii) $\theta_1 = \pi/2,$ $\theta_2 = 4\pi/3,$ $\theta_3 = \pi/3$

(iii) $\theta_1 = -\pi/6,$ $\theta_2 = 2\pi/3,$ $\theta_3 = -\pi/3$

4.2 The wrist of the Cincinnati Milacron T^3 robot is illustrated in fig. 4.4. Choose a suitable home configuration and work out the kinematic transformation matrix of the wrist.

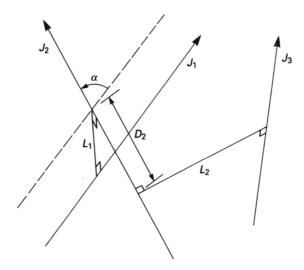

Figure 4.5 Link Length, Twist Angle and Joint Offset

4.3 Design Parameters

For a planar mechanism with hinge joints, we saw that only the lengths of the links are needed to specify the design. For spatial systems the situation is a bit more complicated. To make things as simple as possible we will only look at systems of revolute joints. We then have three kinds of design parameters.

To begin with, consider just two revolute joints, see fig. 4.5. Each joint determines a line in space. Between any pair of lines there is a unique shortest distance, along the line perpendicular to both. This length is called the **link length**. If the lines happen to intersect then the link length is zero. A problem might arise here if the lines are parallel. Then there are many common perpendiculars. However, the distance between the lines along any common perpendicular is always the same, so no ambiguity arises.

Now if we look along the common perpendicular the lines will appear to cross. The angle at which they cross is called the **twist angle**. Another way to think of this is as the angle between the direction vectors along the lines. There are two possible choices for the direction of the line, differing by π radians. But we can make the choice that results in the twist angle α, being in the range $0 \leq \alpha < \pi$. This of course fails if the twist angle is $\frac{\pi}{2}$. In this case we must choose some other criterion to fix the direction of the line.

The final design parameter is the **joint offset**. This is only relevant if we have three or more joints, see fig. 4.5. The line perpendicular to joint 1 and joint 2 meets the axis of joint 2 at some point. Likewise, the common perpendicular to the second and third joint meets the axis of the second joint. The distance between these two points is the joint offset. The joint offset is positive when measured along the direction of the axis. A possible ambiguity is here, if two consecutive axes are parallel; then the joint offset cannot be defined.

By considering fig. 4.6 we can draw up a table of the design parameters for the Puma

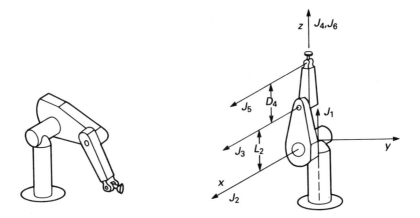

Figure 4.6 The Puma Robot and its Home Position

robot.

Link	Link length	Twist angle	Offset
J_1—J_2	0	$\pi/2$	–
J_2—J_3	L_2	0	(axes parallel)
J_3—J_4	0	$\pi/2$	D_3
J_4—J_5	0	$\pi/2$	D_4
J_5—J_6	0	$\pi/2$	0

These design parameters are also sometimes called the Denavit-Hartenberg parameters. If we include prismatic joints, the offset for such a joint is now a variable. The joint angle, that is the angle between the common perpendiculars, is now fixed, and hence is the new design parameter.

4.4 'A'- Matrices for the Puma Robot

In this section we look at the kinematics of a six axis industrial robot; the Unimate Puma. Many industrial robots follow the design of the Puma: some of the dimensions may be different and also the details of the drive systems, but the basic arrangement of the joints is often the same. Originally the Puma was designed to perform light industrial tasks, especially in the car industry and was designed to match human capabilities in terms of reach and lifting power.

As usual, we begin by choosing a home position for the robot, see fig. 4.6. Next we must find the matrices corresponding to rotations about the joint axes. These are the '**A**'- matrices. If the line has direction vector **v** and **t** is the position vector of a point on the line, then rotations about this line will be given by:-

$$\mathbf{A}(\theta) = \left(\begin{array}{c|c} \mathbf{I} & \mathbf{t} \\ \hline 0 & 1 \end{array}\right) \left(\begin{array}{c|c} \mathbf{R}(\theta, \mathbf{v}) & 0 \\ \hline 0 & 1 \end{array}\right) \left(\begin{array}{c|c} \mathbf{I} & -\mathbf{t} \\ \hline 0 & 1 \end{array}\right)$$

The result is:-

$$\mathbf{A}(\theta) = \left(\begin{array}{c|c} \mathbf{R}(\theta, \mathbf{v}) & (\mathbf{I} - \mathbf{R})\mathbf{t} \\ \hline 0 & 1 \end{array}\right)$$

The axes of the joints in their home positions are given by the following table:-

Joint	v	t
J_1	k	0
J_2	i	0
J_3	i	$L_2\mathbf{k}$
J_4	k	$D_3\mathbf{i}$
J_5	i	$(L_2 + D_4)\mathbf{k}$
J_6	k	$D_3\mathbf{i}$

Now it is a simple matter to find the $\mathbf{A}$-matrices. The first one is simply:-

$$\mathbf{A}_1(\theta_1) = \begin{pmatrix} \cos\theta_1 & -\sin\theta_1 & 0 & 0 \\ \sin\theta_1 & \cos\theta_1 & 0 & 0 \\ 0 & 0 & 1 & 0 \\ 0 & 0 & 0 & 1 \end{pmatrix}$$

The second one is also simple:-

$$\mathbf{A}_2(\theta_2) = \begin{pmatrix} 1 & 0 & 0 & 0 \\ 0 & \cos\theta_2 & -\sin\theta_2 & 0 \\ 0 & \sin\theta_2 & \cos\theta_2 & 0 \\ 0 & 0 & 0 & 1 \end{pmatrix}$$

For $\mathbf{A}_3$ we need to know the term $(\mathbf{I} - \mathbf{R})\mathbf{t}$.

$$\left\{ \begin{pmatrix} 1 & 0 & 0 \\ 0 & 1 & 0 \\ 0 & 0 & 1 \end{pmatrix} - \begin{pmatrix} 1 & 0 & 0 \\ 0 & \cos\theta_3 & -\sin\theta_3 \\ 0 & \sin\theta_3 & \cos\theta_3 \end{pmatrix} \right\} \begin{pmatrix} 0 \\ 0 \\ L_2 \end{pmatrix} = \begin{pmatrix} 0 \\ L_2\sin\theta_3 \\ L_2(1 - \cos\theta_3) \end{pmatrix}$$

So that:-

$$\mathbf{A}_3(\theta_3) = \begin{pmatrix} 1 & 0 & 0 & 0 \\ 0 & \cos\theta_3 & -\sin\theta_3 & L_2\sin\theta_3 \\ 0 & \sin\theta_3 & \cos\theta_3 & L_2(1 - \cos\theta_3) \\ 0 & 0 & 0 & 1 \end{pmatrix}$$

Similarly we have:-

$$\mathbf{A}_4(\theta_4) = \begin{pmatrix} \cos\theta_4 & -\sin\theta_4 & 0 & (1 - \cos\theta_4)D_3 \\ \sin\theta_4 & \cos\theta_4 & 0 & -\sin\theta_4 D_3 \\ 0 & 0 & 1 & 0 \\ 0 & 0 & 0 & 1 \end{pmatrix}$$

and also:-

$$\mathbf{A}_5(\theta_5) = \begin{pmatrix} 1 & 0 & 0 & 0 \\ 0 & \cos\theta_5 & -\sin\theta_5 & (L_2 + D_4)\sin\theta_5 \\ 0 & \sin\theta_3 & \cos\theta_3 & (L_2 + D_4)(1 - \cos\theta_5) \\ 0 & 0 & 0 & 1 \end{pmatrix}$$

Finally we have:-

$$\mathbf{A}_6(\theta_6) = \begin{pmatrix} \cos\theta_6 & -\sin\theta_6 & 0 & (1 - \cos\theta_6)D_3 \\ \sin\theta_6 & \cos\theta_6 & 0 & -\sin\theta_6 D_3 \\ 0 & 0 & 1 & 0 \\ 0 & 0 & 0 & 1 \end{pmatrix}$$

The kinematic matrix is just given by the product:-

$$\mathbf{K}(\theta_1, \theta_2, \theta_3, \theta_4, \theta_5, \theta_6) = \mathbf{A}_1(\theta_1)\mathbf{A}_2(\theta_2)\mathbf{A}_3(\theta_3)\mathbf{A}_4(\theta_4)\mathbf{A}_5(\theta_5)\mathbf{A}_6(\theta_6)$$

As usual the first operation is represented by the rightmost matrix. Here, we may think of performing the rotation about the final joint first, thus leaving the earlier joints fixed. Continuing down the chain we get the above order for the $\mathbf{A}$-matrices. Multiplying out these matrices will give a very large, complicated matrix which we could not fit on a normal page. Luckily, the complicated expressions obtained do not add anything to the understanding of the machine. In a practical situation, we may have to multiply out the matrices once; the resulting expressions would then be programmed into the robot's control system.

Since everything is measured from the home position, when all the angles are zero we should get the identity matrix, and indeed it is not hard to check that:-

$$\mathbf{K}(0,0,0,0,0,0) = \mathbf{I}$$

Again we can use the kinematic matrix to find the position of points with known home co-ordinates. For example, let $\mathbf{p}$ be a point rigidly attached to the gripper, and assume that in the home position $\mathbf{p}$ has co-ordinates $(0, 0, L_2 + D_4)$. Then when the joint angles are:-

$$\theta_1 = \frac{\pi}{2}, \quad \theta_2 = \frac{\pi}{2}, \quad \theta_3 = 0, \quad \theta_4 = -\frac{\pi}{2}, \quad \theta_5 = \frac{\pi}{2}, \quad \theta_6 = \frac{\pi}{2}$$

the new position of the point is given by:-

$$\begin{pmatrix} \mathbf{p}' \\ 1 \end{pmatrix} = \mathbf{K}(\frac{\pi}{2}, \frac{\pi}{2}, 0, -\frac{\pi}{2}, \frac{\pi}{2}, \frac{\pi}{2}) \begin{pmatrix} \mathbf{p} \\ 1 \end{pmatrix}$$

$$= \begin{pmatrix} 1 & 0 & 0 & (L_2 - D_3 + D_4) \\ 0 & 0 & -1 & (L_2 + D_3 + D_4) \\ 0 & 1 & 0 & 0 \\ 0 & 0 & 0 & 1 \end{pmatrix} \begin{pmatrix} 0 \\ 0 \\ (L_2 + D_4) \\ 1 \end{pmatrix} = \begin{pmatrix} (L_2 - D_3 + D_4) \\ D_3 \\ 0 \\ 1 \end{pmatrix}$$

To sum up, the forward kinematics give a mapping from the robot's joint space to its work space. This is expressed by the kinematic transformation matrix; a rigid transformation depending on the joint variables. For open loop robots we have demonstrated a systematic procedure for obtaining this matrix.

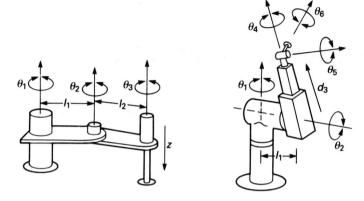

Figure 4.7 The Scara Robot and the Stanford Manipulator

Exercises

4.3   Find the **A** matrices for the Scara robot illustrated in fig. 4.7. (Scara is an acronym for Selective Compliance Arm for Robot Assembly.) Also find the **A** matrices for the 5-R-1-P Stanford manipulator also illustrated in fig. 4.7.

4.4   A point **p** is rigidly attached to the gripper of a Puma type robot. In the home configuration the point has co-ordinates $(D_3, 0, L_2 + D_4)$. Use **A** matrices of the Puma to find its position when the joint angles are:-

$$\theta_1 = 0, \quad \theta_2 = 0, \quad \theta_3 = 0, \quad \theta_4 = \frac{\pi}{3}, \quad \theta_5 = \frac{\pi}{2}, \quad \theta_6 = \frac{\pi}{6}$$

4.5   Another point, **r**, is rigidly attached to the gripper of a Puma type robot. When the joint angles are:-

$$\theta_1 = \frac{\pi}{2}, \quad \theta_2 = \frac{\pi}{2}, \quad \theta_3 = 0, \quad \theta_4 = -\frac{\pi}{2}, \quad \theta_5 = \frac{\pi}{2}, \quad \theta_6 = \frac{\pi}{2}$$

the point **r** has co-ordinates $(L_2 + D_4, D_3 + D_4, 0)$. Find the co-ordinates of the point in the home configuration.

5   Inverse Kinematics

In the last chapter we saw how to derive the kinematics of a serial robot. The position and orientation of any point rigidly attached to the gripper can be found if the joint angles are known. In this section we want to do the reverse. Given the position and orientation of the gripper required, to what angles must the joints be set? This is one of the central problems in robotics, since whenever we specify the motion of the robot's gripper we need to know the corresponding joint motions. Essentially we must solve the following matrix equation:-

$$\mathbf{A}_1(\theta_1)\mathbf{A}_2(\theta_2)\mathbf{A}_3(\theta_3)\mathbf{A}_4(\theta_4)\mathbf{A}_5(\theta_5)\mathbf{A}_6(\theta_6) = \mathbf{K}$$

where $\mathbf{K}$ is the constant matrix which specifies the position and orientation of the gripper. This constitutes a set of highly non-linear equations for the joint angles $\theta_1, \theta_2, \ldots, \theta_6$.

In general, very little is known about solving such equations; even the number of solutions is problematic. For n non-linear equations in n unknowns there may be no solutions at all, one or more discrete solutions or even continuous families of solutions. This contrasts sharply with the case of linear equations, where only a single solution or a linear space of solutions is possible. In the linear case we can look to the determinant of the system to distinguish these cases; for non-linear equations no such test exists.

Things are not quite so bad if there are no helical joints, since the joint angles only appear in the equations as $\cos\theta_i$ or $\sin\theta_i$. Now if we use these as our variables the equations are **algebraic**. That is, they are only polynomials in the variables $\cos\theta_i$ and $\sin\theta_i$. So if we solve for these variables it is a simple matter to find the joint angles; θ_i. However, we have actually doubled the number of variables in the equations but we must also consider the relations between the new variables. This means we must include the equations:-

$$\cos^2\theta_i + \sin^2\theta_i = 1$$

in our non-linear system.

There is another technique used to make the equations algebraic: to write the equations in terms of 'tan half angles'; that is, to make the substitutions:-

$$\cos\theta_i = \frac{1 - t_i^2}{1 + t_i^2} \qquad \sin\theta_i = \frac{2t_i}{1 + t_i^2}$$

where $t_i = \tan(\theta_i/2)$. The only disadvantage of this approach is that it fails when $\theta_i = \pi$.

Algebraic equations have nice properties. For example a polynomial equation in one variable has as many solutions (roots) as the degree of the polynomial. This is familiar from elementary algebra, and also we recall that the roots must be counted properly; repeated roots and complex roots must be accounted for. There is a generalization of this to systems of polynomial equations in several variables. If we have n equations of degree $d_1, d_2, \ldots, d_n$ in n unknowns then, in general, we get $d_1 \times d_2 \times \ldots \times d_n$ solutions. However, there are exceptional circumstances when there is an infinite family of solutions.

So, for example, consider two quadratics in two variables. Quadratics in two variables are just conic curves; ellipses, parabolas and hyperbolas. Their degree is two, so two of them should intersect in $2 \times 2 = 4$ points. Some of these intersections may be complex; they will occur in complex conjugate pairs if the coefficients of the equations are real. Hence there may be no real intersections at all. Singular solutions are also possible. They correspond to repeated roots in the one variable case, and occur when the curves intersect and have the same tangent at the intersection. See fig. 5.1.

5.1   The Planar Manipulator

To get back to the problem of inverse kinematics let us look at a simple example. The planar manipulator exhibits all the possibilities that can arise. Consider the position after just two links, see fig. 5.2. The kinematic equations for the end point are:-

$$x = l_1 \cos\theta_1 + l_2 \cos(\theta_1 + \theta_2)$$
$$y = l_1 \sin\theta_1 + l_2 \sin(\theta_1 + \theta_2)$$

Given x and y we must find $\cos\theta_1, \sin\theta_1, \cos\theta_2$ and $\sin\theta_2$. The above equations are in fact quadratic, since we can use the trigonometric formulas to write:-

$$\cos(\theta_1 + \theta_2) = \cos\theta_1 \cos\theta_2 - \sin\theta_1 \sin\theta_2 \qquad \sin(\theta_1 + \theta_2) = \sin\theta_1 \cos\theta_2 + \cos\theta_1 \sin\theta_2$$

Then together with the identities satisfied by the sine and cosine functions, these give us four quadratic equations in four unknowns:-

$$x = l_1 \cos\theta_1 + l_2 \cos\theta_1 \cos\theta_2 - l_2 \sin\theta_1 \sin\theta_2 \qquad (A)$$
$$y = l_1 \sin\theta_1 + l_2 \sin\theta_1 \cos\theta_2 + l_2 \cos\theta_1 \sin\theta_2 \qquad (B)$$
$$1 = \cos^2\theta_1 + \sin^2\theta_1 \qquad (C)$$
$$1 = \cos^2\theta_2 + \sin^2\theta_2 \qquad (D)$$

So we might expect $2 \times 2 \times 2 \times 2 = 16$ solutions: in fact only 2 arise. The discrepancy is accounted for by four singular complex solutions at 'infinity'.

To solve this system we square equation (A) and add it to the square of (B):-

$$(x^2 + y^2) = l_1^2(\cos^2\theta_1 + \sin^2\theta_1) + l_2^2(\cos^2\theta_1 + \sin^2\theta_1)\cos^2\theta_2$$
$$+ l_2^2(\cos^2\theta_1 + \sin^2\theta_1)\sin^2\theta_2 + 2l_1l_2(\cos^2\theta_1 + \sin^2\theta_1)\cos\theta_2$$

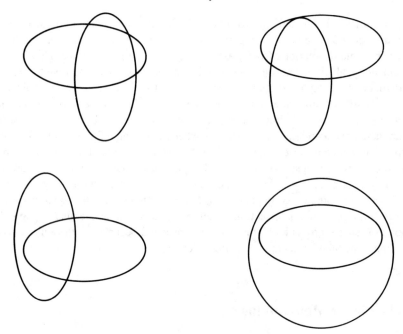

Figure 5.1 Some Possible Intersections of Two Conics

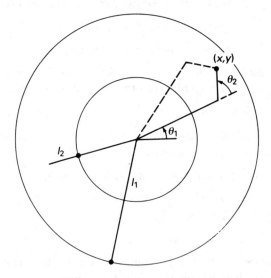

Figure 5.2 The Planar Manipulator; Postures and Work Space

Then using (C) and (D) to simplify we obtain:-

$$(x^2 + y^2) = l_1^2 + l_2^2 + 2l_1 l_2 \cos \theta_2$$

In fact this is just the cosine rule from trigonometry. So the solution for $\cos \theta_2$ is just:-

$$\cos \theta_2 = \frac{1}{2l_1 l_2} \{(x^2 + y^2) - (l_1^2 + l_2^2)\} = \lambda$$

We abbreviate this to λ since it will occur frequently. Hence, by (D) $\sin \theta_2$ is:-

$$\sin \theta_2 = \pm (1 - \lambda^2)^{\frac{1}{2}}$$

that is, there are two possible solutions. We will look at this in more detail in a moment. First let us find $\cos \theta_1$ and $\sin \theta_1$. The simplest way to do this is to form the two equations:-

$$
\begin{aligned}
(A) \cos \theta_1 + (B) \sin \theta_1 &\equiv x \cos \theta_1 + y \sin \theta_1 &= l_1 + l_2 \cos \theta_2 \\
&&= l_1 + l_2 \lambda
\end{aligned}
$$

$$
\begin{aligned}
-(A) \sin \theta_1 + (B) \cos \theta_1 &\equiv -x \sin \theta_1 + y \cos \theta_1 &= l_2 \sin \theta_2 \\
&&= \pm l_2 (1 - \lambda^2)^{\frac{1}{2}}
\end{aligned}
$$

Again, the relation (C) has been used to simplify the above. Now we have two simultaneous linear equations which are easily solved. So we have found explicit equations for the sines and cosines of the angles in terms of the design parameters and the position of the manipulator's end-effector. In fact there are two solutions, corresponding to the upper and lower sign choices. These equations are the inverse kinematic relations for the manipulator:-

$$\cos \theta_1 = \frac{1}{(x^2 + y^2)} \{x(l_1 + l_2 \lambda) \pm y l_2 (1 - \lambda^2)^{\frac{1}{2}}\}$$

$$\sin \theta_1 = \frac{1}{(x^2 + y^2)} \{\mp x l_2 (1 - \lambda^2)^{\frac{1}{2}} + y(l_1 + l_2 \lambda)\}$$

$$\cos \theta_2 = \frac{1}{2l_1 l_2} \{(x^2 + y^2) - (l_1^2 + l_2^2)\} = \lambda$$

$$\sin \theta_2 = \pm (1 - \lambda^2)^{\frac{1}{2}}$$

5.2 Postures

For the planar manipulator of the previous section there are generally two solutions for the inverse kinematics. They arise from the sign of the term $\sin \theta_2$: physically this corresponds to the fact that there are two ways of reaching any point in the plane, see fig. 5.2. These two configurations of the manipulator are called **postures**; one is referred to as 'elbow up', the other as 'elbow down'. However, not every point (x, y) has two postures. There is only one solution for $\sin \theta_2$ if $\sin \theta_2 = 0$; that is when $\lambda = \pm 1$, which corresponds to $\theta_2 = 0$ or

π. The points in the plane determined by these values are given by:-

$$\cos\theta_2 = 1; \quad \text{gives} \quad (x^2 + y^2) = (l_1 + l_2)^2$$
$$\cos\theta_2 = -1; \quad \text{gives} \quad (x^2 + y^2) = (l_1 - l_2)^2$$

These are the equations of two concentric circles, on one the arm is at full stretch, while to reach the other the arm must double back on itself, see fig. 5.2.

Beyond the outer circle and inside the smaller one, the solutions for $\sin\theta_2$ become complex and it is clear that we cannot reach such points with a real arm. The annular region is the projection of the robot's **work space** onto the plane, see section 3.3. It is the space that the robot can reach and work in. The work space of any robot is always bounded by curves or surfaces on which the number of postures is different from the body of the work space. Such points are called **singular points**; however, singular points may also occur in the interior of the work space. A better characterization of singular points is points where the robot loses one or more degrees-of-freedom. In the case of the planar manipulator it is easy to see that on the boundary of the work space the arm has no freedom to move in a radial direction.

So far we have said nothing about the design parameters l_1 and l_2. In fact the relative sizes of the links do not affect the number of postures, except in the very special case that $l_1 = l_2$. In this case there are still generally two postures for every point in the work space, but the inner boundary has now shrunk to a point; the origin. If we try to place the tip of the manipulator at $x = 0$, $y = 0$, then certainly we must have $\cos\theta_2 = -1$ and $\sin\theta_2 = 0$, but our method for finding θ_1 breaks down. It is quite clear though that there is no restriction on θ_1, so instead of one or two postures, this point in the work space has a whole circle of postures. This kind of singularity, with a continuous family of postures, is particularly difficult to deal with when it comes to controlling the robot. Unfortunately, all six axis robots that have been designed or built have such singularities in their work space. It is not known if this can be avoided.

5.3   The 3-R Wrist

As we saw in section 4.2 the kinematic equations of the 3-R wrist can be written in terms of the Euler angles as:-

$$\psi = \theta_1, \qquad \theta = \theta_2, \qquad \phi = \theta_3$$

so there does not seem to be any problem about the inverse kinematics. However, we must remember that the Euler angles have a limited range whilst the joint angles can range over a full circle, 0 to 2π; at least in theory. The kinematic matrix, as we saw in section 4.2, is given by:-

K =

$$\begin{pmatrix} \cos\theta_1\cos\theta_2\cos\theta_3 - \sin\theta_1\sin\theta_3 & -\cos\theta_1\cos\theta_2\sin\theta_3 - \sin\theta_1\cos\theta_3 & \cos\theta_1\sin\theta_2 \\ \sin\theta_1\cos\theta_2\cos\theta_3 + \cos\theta_1\sin\theta_3 & -\sin\theta_1\cos\theta_2\sin\theta_3 + \cos\theta_1\cos\theta_3 & \sin\theta_1\sin\theta_2 \\ -\sin\theta_2\cos\theta_3 & \sin\theta_2\sin\theta_3 & \cos\theta_2 \end{pmatrix}$$

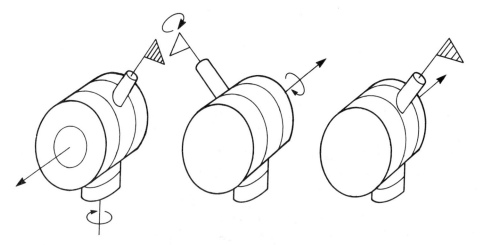

Figure 5.3 Flip and No Flip Postures of the 3-R Wrist

Notice that we get the same matrix if we make the substitutions:-

$$\theta_1' = \pi + \theta_1, \qquad \theta_2' = 2\pi - \theta_2, \qquad \theta_3' = \pi + \theta_3$$

This is because we can use the usual trigonometric relations to give:-

$$
\begin{aligned}
\cos\theta_1' &= -\cos\theta_1, & \cos\theta_2' &= \cos\theta_2, & \cos\theta_3' &= -\cos\theta_3 \\
\sin\theta_1' &= -\sin\theta_1, & \sin\theta_2' &= -\sin\theta_2, & \sin\theta_3' &= -\sin\theta_3
\end{aligned}
$$

So we have two postures, that is two possible solutions for the joint angles given a 3×3 rotation matrix. In terms of the Euler angles these solutions can be written as:-

$$
\begin{aligned}
\theta_1 &= \psi \quad \text{or} \quad \pi + \psi \\
\theta_2 &= \theta \quad \text{or} \quad 2\pi - \theta \\
\theta_3 &= \phi \quad \text{or} \quad \pi - \phi
\end{aligned}
$$

These two postures have been given the names 'flip' and 'no flip'. Fig. 5.3 shows how to change from one posture to the other.

The above results have not been derived in a very systematic way: it is difficult to see whether or not there are other solutions. We can repeat the analysis more efficiently by looking at the effect of the kinematic matrix on two points. Suppose $\mathbf{a} = (0, 0, 1)^T$ and $\mathbf{b} = (1, 0, 0)^T$ are the home positions of two points rigidly attached to the gripper. Then rotating about the three wrist joints will take the points to:-

$$\mathbf{a}' = \mathbf{K}(\theta_1, \theta_2, \theta_3)\mathbf{a} \qquad \text{and} \qquad \mathbf{b}' = \mathbf{K}(\theta_1, \theta_2, \theta_3)\mathbf{b}$$

The co-ordinates of these new points are easily calculated:-

$$\mathbf{a'} = \begin{pmatrix} x_a \\ y_a \\ z_a \end{pmatrix} = \begin{pmatrix} \cos\theta_1 \sin\theta_2 \\ \sin\theta_1 \sin\theta_2 \\ \cos\theta_2 \end{pmatrix}$$

$$\mathbf{b'} = \begin{pmatrix} x_b \\ y_b \\ z_b \end{pmatrix} = \begin{pmatrix} \cos\theta_1 \cos\theta_2 \cos\theta_3 - \sin\theta_1 \sin\theta_3 \\ \sin\theta_1 \cos\theta_2 \cos\theta_3 + \cos\theta_1 \sin\theta_3 \\ -\sin\theta_2 \cos\theta_3 \end{pmatrix}$$

The points have been chosen to simplify the calculations as far as possible. For example, the new position vectors of the points are just the first and third rows of the matrix $\mathbf{K}$.

From the first point we see that $\cos\theta_2 = z_a$ and hence we can find the sine of the angle $\sin\theta_2 = \pm\sqrt{(1-z_a^2)}$. The sine and cosine of the first joint angle can now be found from the x and y co-ordinates of this point; $\cos\theta_1 = \pm x_a/\sqrt{(1-z_a^2)}$ and $\sin\theta_1 = \pm y_a/\sqrt{(1-z_a^2)}$. To find the third joint angle we must look at the second point, then; $\cos\theta_3 = \mp z_b/\sqrt{(1-z_a^2)}$. The sine of θ_3 can be found from x_b and y_b:-

$$\begin{aligned} \sin\theta_3 &= y_b\cos\theta_1 - x_b\sin\theta_1 \\ &= \pm\left\{ \frac{x_a y_b - x_b y_a}{\sqrt{(1-z_a^2)}} \right\} \end{aligned}$$

To summarize, the inverse kinematics of the 3-R wrist, in terms of the positions of the two points, is given by:-

$$\begin{aligned} \cos\theta_1 &= \pm\frac{x_a}{\sqrt{(1-z_a^2)}}, & \sin\theta_1 &= \pm\frac{y_a}{\sqrt{(1-z_a^2)}} \\ \cos\theta_2 &= z_a, & \sin\theta_2 &= \pm\sqrt{(1-z_a^2)} \\ \cos\theta_3 &= \mp\frac{z_b}{\sqrt{(1-z_a^2)}}, & \sin\theta_3 &= \pm\left\{\frac{x_a y_b - x_b y_a}{\sqrt{(1-z_a^2)}}\right\} \end{aligned}$$

The two postures are distinguished by the sign of $\sin\theta_2$. Notice that this analysis also tells us where the number of postures is different from two, since, if $\sin\theta_2 = 0$, we cannot divide by this factor and the above analysis fails. In fact these two points with $\theta_2 = 0$ or π, are singular points each with an infinite number of postures. At these points the first and last axes coincide, thus the final link can be held fixed while the second joint rotates perpendicular to the first and last joints; see also section 2.5.

Exercises

5.1   A planar manipulator has link lengths $l_1 = 2$ and $l_2 = 1$ in some units. Use the inverse kinematic equations to find the joint angles which will place the end point at the following positions:-

(i) $x = (\sqrt{3} + \frac{1}{2})$, $y = 1 + \frac{\sqrt{3}}{2}$

(ii) $x = 2,$ $y = 1 + \sqrt{3}$

(iii) $x = \sqrt{2},$ $y = 1 + \sqrt{2}$

5.2   The two points $(0, 0, 1)$ and $(1, 0, 0)$ are rigidly attached to the gripper of a 3-R wrist. Use the inverse kinematic equations derived in the text to find the joint angles when these points have the co-ordinates:-

(i) $(\frac{1}{2}, 0, \frac{\sqrt{3}}{2})$ and $(\frac{3}{4}, -\frac{1}{2}, -\frac{\sqrt{3}}{4})$

(ii) $(\frac{3}{4}, \frac{\sqrt{3}}{4}, \frac{1}{2})$ and $(\frac{\sqrt{3}}{4}, \frac{1}{4}, -\frac{\sqrt{3}}{2})$

(iii) $(\frac{3}{4}, \frac{\sqrt{3}}{4}, \frac{1}{2})$ and $(\frac{1}{8}, -\frac{\sqrt{3}}{8}, -\frac{\sqrt{3}}{4})$

5.3   Work out the inverse kinematic relations for the three joint planar manipulator studied in section 4.1. If the links' lengths are $l_1 = 2, l_2 = 1$ and $l_3 = 1$ in some system of units, find the possible joint angles which result in the end point having co-ordinates $x = 0.5, y = 3.0$ and output angle $\Phi = 2\pi/3$ radians.

5.4   Work out the inverse kinematics of the 3-R wrist in terms of the positions of two points rigidly attached to the gripper and where the home co-ordinates are $(0, 0, 1)$ and $(0, 1, 0)$.

5.4   The First Three Joints of the Puma

Now we are in a position to calculate the inverse kinematics for the Puma arm. This is possible because the first three joints of the Puma are almost a planar manipulator while the last three are a 3-R wrist. Hence, the problem can be split into two easier pieces. We will express the inverse kinematics in terms of the components of three points rigidly attached to the gripper. In the home position these points will have co-ordinates:-

$$\mathbf{p}_a = \begin{pmatrix} D_3 \\ 0 \\ L_2 + D_4 + 1 \end{pmatrix}, \qquad \mathbf{p}_b = \begin{pmatrix} D_3 + 1 \\ 0 \\ L_2 + D_4 \end{pmatrix}, \qquad \mathbf{p}_c = \begin{pmatrix} D_3 \\ 0 \\ L_2 + D_4 \end{pmatrix}$$

These points have been chosen to make things easy, for example $\mathbf{p}_c$ is the position of the wrist centre. Hence, only movements about the first three joints will affect the position of $\mathbf{p}_c$. Moreover, if we know the position of the wrist centre we can find solutions for the first three joints.

The forward kinematics, or just a consideration of the geometry, gives:-

$$\begin{aligned} x_c &= D_3 \cos\theta_1 + L_2 \sin\theta_1 \sin\theta_2 + D_4 \sin\theta_1 \sin(\theta_2 + \theta_3) \\ y_c &= D_3 \sin\theta_1 - L_2 \cos\theta_1 \sin\theta_2 - D_4 \cos\theta_1 \sin(\theta_2 + \theta_3) \\ z_c &= L_2 \cos\theta_2 + D_4 \cos(\theta_2 + \theta_3) \end{aligned}$$

See fig. 5.4.

Since the second and third joint axes are parallel they behave like a planar manipulator. The first joint simply allows rotation of the plane. So we could write these relations in

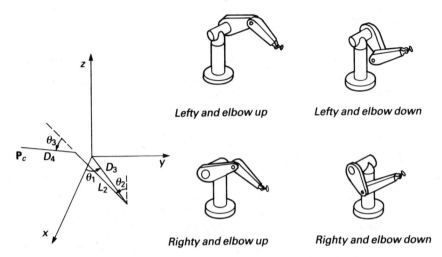

Figure 5.4 The First Three Joints of the Puma and the Possible Postures

terms of the kinematics of a planar manipulator:-

$$x_c = D_3 \cos\theta_1 - \sin\theta_1 r_y$$
$$y_c = D_3 \sin\theta_1 + \cos\theta_1 r_y$$
$$z_c = r_z$$

where $r_y = -L_2 \sin\theta_2 - D_4 \sin(\theta_2 + \theta_3)$ and $r_z = L_2 \cos\theta_2 + D_4 \cos(\theta_2 + \theta_3)$ can be thought of as the forward kinematics of a planar manipulator. The first two equations can be rearranged to give:-

$$D_3 = x_c \cos\theta_1 + y_c \sin\theta_1$$
$$r_y = y_c \cos\theta_1 - x_c \sin\theta_1$$

The effect is the same as multiplying by the inverse of $\mathbf{A}_1(\theta_1)$. The first of these new equations is linear in the sine and cosine of θ_1, so we can use it to eliminate $\sin\theta_1$ from the quadratic equation $\cos^2\theta_1 + \sin^2\theta_1 = 1$:-

$$(x_c^2 + y_c^2)\cos^2\theta_1 - 2x_c D_3 \cos\theta_1 + (D_3^2 - y_c^2) = 0$$

Solving this for $\cos\theta_1$ we get:-

$$\cos\theta_1 = \frac{D_3 x_c \pm y_c \sqrt{x_c^2 + y_c^2 - D_3^2}}{(x_c^2 + y_c^2)}$$

using the standard solution for a quadratic. The sine is given by substituting this back in the linear equation:-

$$\sin\theta_1 = \frac{1}{y_c}\left\{ D_3 - x_c \frac{D_3 x_c \pm y_c \sqrt{x_c^2 + y_c^2 - D_3^2}}{(x_c^2 + y_c^2)} \right\}$$

These expressions are going to get very complicated, so to simplify things as much as possible we will write the solutions in terms of the ones we have already found. So it is clear from the above that we could write everything explicitly in terms of the co-ordinates of the points, but we will content ourselves with implicit relations.

Evidently we have two possible solutions for θ_1, depending on the sign of the square root. Each will result in a different posture as we shall see later. Next we use the inverse kinematics of the planar manipulator to solve for θ_2 and θ_3:-

$$\cos\theta_3 = \frac{1}{2L_2D_4}\{(r_y^2 + r_z^2) - (L_2^2 + D_4^2)\}$$

$$\sin\theta_3 = \pm\sqrt{1 - \cos^2\theta_3}$$

This introduces a second ambiguity in sign:-

$$\cos\theta_2 = \frac{1}{(r_y^2 + r_z^2)}\{r_z(L_2 + D_4\cos\theta_3) - r_yD_4\sin\theta_3\}$$

$$\sin\theta_2 = \frac{1}{(r_y^2 + r_z^2)}\{-r_y(L_2 + D_4\cos\theta_3) + r_zD_4\sin\theta_3\}$$

These results were simply obtained by substituting $x = r_z, y = -r_y$ in the results of section 5.1.

The two possible sign choices are independent of each other, so there are four possible solutions, and hence four postures. For the Puma these have the cute names 'elbow up' or 'elbow down' depending on the choice of the sign of $\sin\theta_3$, and 'righty' or 'lefty' depending on which sign of $\sqrt{x_c^2 + y_c^2 - D_3^2}$ is chosen. Notice that a little rearrangement gives $\sqrt{x_c^2 + y_c^2 - D_3^2} = (y_c\cos\theta_1 - x_c\sin\theta_1)$. Hence given a set of joint angles, we can tell which posture the robot is in by looking at the sign of these two functions. Including the two possible postures for the wrist, the Puma has eight different postures in all, see fig. 5.4.

5.5   The Last Three Joints of the Puma

Most of the hard work here has been done in section 5.3. The only difference is the effect of the first three joints. Remember we are trying to solve the equations:-

$$\mathbf{A}_1(\theta_1)\mathbf{A}_2(\theta_2)\mathbf{A}_3(\theta_3)\mathbf{A}_4(\theta_4)\mathbf{A}_5(\theta_5)\mathbf{A}_6(\theta_6)\mathbf{p} = \mathbf{p}'$$

where $\mathbf{p} = \mathbf{p}_a$ or $\mathbf{p}_b$. This equation can be rearranged to give:-

$$\mathbf{A}_4(\theta_4)\mathbf{A}_5(\theta_5)\mathbf{A}_6(\theta_6)\mathbf{p} = \mathbf{A}_3^{-1}(\theta_3)\mathbf{A}_2^{-1}(\theta_2)\mathbf{A}_1^{-1}(\theta_1)\mathbf{p}' \qquad (**)$$

Now, the right-hand side of the above equation is, in principle, known. Also we have chosen the points $\mathbf{p}_a$ and $\mathbf{p}_b$ to be in the same relation to the wrist centre $\mathbf{p}_c$, as the points

a and **b** were to the origin in section 5.3. In fact we can write:-

$$\mathbf{p}_a = \mathbf{p}_c + \mathbf{a} \qquad \text{and} \qquad \mathbf{p}_b = \mathbf{p}_c + \mathbf{b}$$

Equation (∗∗) above can now be written as:-

$$\mathbf{K}(\theta_4, \theta_5, \theta_6)\mathbf{a} + \mathbf{p}_c = \mathbf{R}(-\theta_3, \mathbf{i})\mathbf{R}(-\theta_2, \mathbf{i})\mathbf{R}(-\theta_1, \mathbf{k})\mathbf{p}'_a - \mathbf{R}(-\theta_3, \mathbf{i})\mathbf{v}$$

Here **K** is the kinematic matrix of the 3-R wrist, as in section 5.3. Since $\mathbf{p}_c$ lies on all the axes of the wrist it is not affected by the kinematics of the wrist. On the right-hand side of the equation we have the inverses of the rotations about the first three joints; the term in **v** results from the translation part of **A** $_3$. We also get a similar equation for **b**. Now it is possible to rearrange the above equation into the form we solved in section 5.3:-

$$\mathbf{K}(\theta_4, \theta_5, \theta_6)\mathbf{a} = \boldsymbol{\alpha} \qquad \mathbf{K}(\theta_4, \theta_5, \theta_6)\mathbf{b} = \boldsymbol{\beta}$$

The vector $\boldsymbol{\alpha}$ is given by:-

$$\boldsymbol{\alpha} = \begin{pmatrix} x_a \cos\theta_1 + y_a \sin\theta_1 - D_3 \\ x_a \sin\theta_1 \cos(\theta_2 + \theta_3) + y_a \cos\theta_1 \cos(\theta_2 + \theta_3) + z_a \sin(\theta_2 + \theta_3) - L_2 \sin\theta_3 \\ x_a \sin\theta_1 \sin(\theta_2 + \theta_3) - y_a \cos\theta_1 \sin(\theta_2 + \theta_3) + z_a \cos(\theta_2 + \theta_3) - L_2 \cos\theta_3 - D_4 \end{pmatrix}$$

The vector $\boldsymbol{\beta}$ has a similar expression, with x_a, y_a and z_a replaced by x_b, y_b and z_b. It is now just a matter of substituting these expressions into the solutions we have already found for the 3-R wrist. Although tedious, the procedure is straightforward. The result is not particularly instructive, but would be necessary for the robot's control system.

5.6 Inverse Kinematics of the Puma

We may summarize the results of the last two sections in the following page of equations.

$$\cos\theta_1 = \{D_3 x_c \pm y_c \sqrt{x_c^2 + y_c^2 - D_3^2}\}/(x_c^2 + y_c^2)$$

$$\sin\theta_1 = \{D_3 - x_c \cos\theta_1\}/y_c$$

$$r_y = y_c \cos\theta_1 - x_c \sin\theta_1$$

$$r_z = z_c$$

$$\cos\theta_3 = \{(r_y^2 + r_z^2) - (L_2^2 + D_4^2)\}/2L_2 D_4$$

$$\sin\theta_3 = \pm\sqrt{1 - \cos^2\theta_3}$$

$$\cos\theta_2 = \{r_z(L_2 + D_4 \cos\theta_3) - r_y D_4 \sin\theta_3\}/(r_y^2 + r_z^2)$$

$$\sin\theta_2 = \{-r_y(L_2 + D_4 \cos\theta_3) + r_z D_4 \sin\theta_3\}/(r_y^2 + r_z^2)$$

$$z_\alpha = x_a \sin\theta_1 \sin(\theta_2 + \theta_3) - y_a \cos\theta_1 \sin(\theta_2 + \theta_3) + z_a \cos(\theta_2 + \theta_3)$$
$$- L_2 \cos\theta_3 - D_4$$

$$\cos\theta_5 = z_\alpha$$

$$\sin \theta_5 = \pm \sqrt{1 - \cos^2 \theta_5}$$

$$x_\alpha = x_a \cos \theta_1 + y_a \sin \theta_1 - D_3$$

$$y_\alpha = -x_a \sin \theta_1 \cos(\theta_2 + \theta_3) + y_a \cos \theta_1 \cos(\theta_2 + \theta_3) + z_a \sin(\theta_2 + \theta_3)$$
$$- L_2 \sin \theta_3$$

$$\cos \theta_4 = x_\alpha / \sin \theta_5$$

$$\sin \theta_4 = y_\alpha / \sin \theta_5$$

$$z_\beta = x_b \sin \theta_1 \sin(\theta_2 + \theta_3) - y_b \cos \theta_1 \sin(\theta_2 + \theta_3) + z_b \cos(\theta_2 + \theta_3)$$
$$- L_2 \cos \theta_3 - D_4$$

$$y_\beta = -x_b \sin \theta_1 \cos(\theta_2 + \theta_3) + y_b \cos \theta_1 \cos(\theta_2 + \theta_3) + z_b \sin(\theta_2 + \theta_3)$$
$$- L_2 \sin \theta_3$$

$$\cos \theta_6 = -z_\beta / \sin \theta_5$$

$$\sin \theta_6 = (y_\beta - \sin \theta_4 \cos \theta_5 \cos \theta_6) / \cos \theta_4$$

Although this looks horribly complicated, at least a solution is possible. If we had chosen the joints arbitrarily then the tricks we used would not have worked. For such general cases analytic solutions are not possible, and usually numerical techniques have to be used. This can be a problem if the number of postures is not known; most numerical methods will only give a single solution. For the general six joint serial robot the number of postures is believed to be sixteen. How the number of postures changes as the design parameters are altered can only be guessed, at present. This is why there are so few different designs of robots: only the ones with analytic solutions for the inverse kinematics tend to be used. However, the range of designs for which the last three joint axes intersect in a common point do always have an analytic solution.

5.7 Parallel Manipulators

The inverse kinematics of parallel manipulators like the Stewart platform are surprisingly straightforward. In fact, it is the forward kinematics which are hard here, which is why we have not studied them earlier. To keep things simple we will only look at a planar parallel manipulator; see fig. 5.5. The mechanism has three sliding joints attached to both the ground link and the movable link by hinge joints. This means that the movable link has three degrees-of-freedom, the correct amount for a planar manipulator.

Let us choose our co-ordinates so that the hinges on the ground link are at the points $\mathbf{p}_1 = (0, 0)$ and $\mathbf{p}_2 = (1, 0)$. Again for convenience, let us choose the two points which determine the position and orientation of the movable link to be the centres of the hinges attached to that link. We denote them $\mathbf{a}$ and $\mathbf{b}$. Now the inverse kinematic problem is to find the joint variables given the two points $\mathbf{a}$ and $\mathbf{b}$: in this case the joint variables are the lengths of the three sliding joints. Elementary geometry gives these lengths as:-

$$d_{a1} = |\mathbf{a} - \mathbf{p}_1|, \qquad d_{b1} = |\mathbf{b} - \mathbf{p}_1|, \qquad d_{b2} = |\mathbf{b} - \mathbf{p}_2|$$

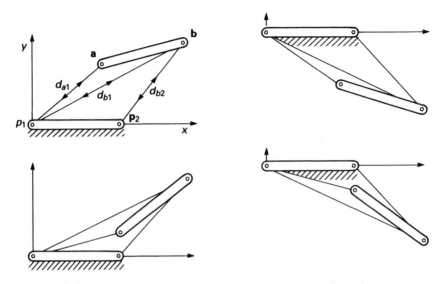

Figure 5.5 The Planar Parallel Manipulator and its Conformations

The inverse kinematics gives a unique solution, so there are no complications with postures here.

As mentioned above, the forward kinematics is more complicated. Given the joint variables we seek the position and orientation of the movable link. From a consideration of the geometry we get four equations:-

$$
\begin{aligned}
(\mathbf{a} - \mathbf{p_1}) \cdot (\mathbf{a} - \mathbf{p_1}) &= d_{a1}^2 \\
(\mathbf{b} - \mathbf{p_1}) \cdot (\mathbf{b} - \mathbf{p_1}) &= d_{b1}^2 \\
(\mathbf{b} - \mathbf{p_2}) \cdot (\mathbf{b} - \mathbf{p_2}) &= d_{b2}^2 \\
(\mathbf{a} - \mathbf{b}) \cdot (\mathbf{a} - \mathbf{b}) &= 1
\end{aligned}
$$

The middle two of these equations can be expanded to:-

$$
(b_x^2 + b_y^2) = d_{b1}^2 \qquad \text{and} \qquad (b_x^2 + b_y^2) - 2b_x + 1 = d_{b2}^2
$$

So we can immediately solve for b_x:-

$$
b_x = \frac{1}{2}(1 + d_{b1}^2 - d_{b2}^2)
$$

Hence we get two solutions for b_y:-

$$
b_y = \pm\sqrt{d_{b1}^2 - b_x^2}
$$

The first and last equations can now be written as:-

$$
(a_x^2 + a_y^2) = d_{a1}^2 \qquad \text{and} \qquad a_x b_x + a_y b_y = d_{a1}^2 + d_{b1}^2 - 1
$$

Since b_x and b_y are now known, the second of these equations can be considered as linear. So, assuming that $b_y \neq 0$ we can substitute for a_y in the quadratic equation to give:-

$$d_{b1}^2 a_x^2 - 2(d_{a1}^2 + d_{b1}^2 - 1)b_x b_y a_x + (d_{a1}^4 + d_{b1}^4 - 3d_{a1}^2 - 2d_{b1}^2 + 2d_{a1}^2 d_{b1}^2 - 1) = 0$$

The familiar solution for a quadratic equation in one variable can now be applied:-

$$a_x = \frac{-B \pm \sqrt{B^2 - 4AC}}{2A}$$

where

$$
\begin{aligned}
A &= d_{b1}^2 \\
B &= -2(d_{a1}^2 + d_{b1}^2)b_x b_y \\
C &= d_{a1}^4 + d_{b1}^4 - 3d_{a1}^2 - 2d_{b1}^2 + 2d_{a1}^2 d_{b1}^2 - 1
\end{aligned}
$$

Finally, we recover a_y from the linear equation:-

$$a_y = \frac{1}{b_y}(d_{a1}^2 + d_{b1}^2 - 1) - a_x b_x$$

Notice that we get four possible solutions in general, called **conformations**. Different conformations have the same values for the joint variables but correspond to different positions and orientations of the end-effector. This is exactly the opposite way around to serial manipulators, but we will get the same kinds of phenomena as with a serial manipulator. The number of conformations will in general be four but will be less at singular positions.

The Stewart platform is rather harder than this example, usually the forward kinematics is done numerically. For a general manipulator, which is neither serial nor parallel, both the forward and inverse kinematics will be hard. Both will involve the solution of sets of algebraic equations and we should expect both multiple postures and conformations to be present.

Exercises

5.5 The wrist centre of a Puma robot is located at the following positions:-

$$\text{(i) } \mathbf{p}_c = \begin{pmatrix} 5/\sqrt{2} \\ -3/\sqrt{2} \\ -4 \end{pmatrix}, \qquad \text{(ii) } \mathbf{p}_c = \begin{pmatrix} -(2 + \frac{3}{\sqrt{2}}) \\ -(2 + \frac{5}{\sqrt{2}}) \\ \frac{4}{\sqrt{2}} \end{pmatrix}.$$

Use the inverse kinematic relations given in the previous sections to find the possible settings for the first three joint angles in each case. Take $L_2 = 4, D_3 = 1$ and $D_4 = 4$.

5.6 Find the inverse kinematic relations for the first three joints of the Stanford manipulator, see fig 4.7. In particular find the joint variables $(\theta_1, \theta_2, d_3)$ in terms of the co-ordinates of the wrist centre (x_c, y_c, z_c). How many postures does such a manipulator have?

5.7   Consider the planar parallel manipulator introduced in section 5.7. Let its home position be when $\mathbf{a} = (0, 1)$ and $\mathbf{b} = (1, 1)$. Suppose the movable link undergoes a rigid transformation given by the matrix:-

$$\begin{pmatrix} \cos\theta & -\sin\theta & t_x \\ \sin\theta & \cos\theta & t_y \\ 0 & 0 & 1 \end{pmatrix}$$

Find the lengths of the sliding joint as functions of t_x, t_y and θ.

6 Jacobians

6.1 Linearized Kinematics

In previous chapters we have seen how kinematics relates the joint angles to the position and orientation of the robot's end-effector. This means that, for a serial robot, we may think of the forward kinematics as a mapping from joint space to the space of rigid body motions. The image of this mapping is the work space of the robot. In general, the work space will be only a subspace of the space of all rigid body motions; it consists of all positions and orientations reachable by the robot's end-effector. As we have already mentioned, we can only put local co-ordinates or parameterisations on the space of rigid body motions[†].

We can also consider mappings associated with particular points; note that the image of such a map is sometimes called the work space of the robot: it is the space of point reachable by some point on the end-effector. Consider, for example the wrist centre of a Puma robot:-

$$\mathbf{K} : (\theta_1, \theta_2, \theta_3, \theta_4, \theta_5, \theta_6) \longrightarrow (x'_c, y'_c, z'_c)$$

The map is given explicitly in terms of the $\mathbf{A}$-matrices:-

$$\begin{pmatrix} x'_c \\ y'_c \\ z'_c \\ 1 \end{pmatrix} = \mathbf{A}_1(\theta_1)\mathbf{A}_2(\theta_2)\mathbf{A}_3(\theta_3)\mathbf{A}_4(\theta_4)\mathbf{A}_5(\theta_5)\mathbf{A}_6(\theta_6) \begin{pmatrix} x_c \\ y_c \\ z_c \\ 1 \end{pmatrix}$$

Here, (x_c, y_c, z_c) are the home co-ordinates of the wrist centre. In other words we have three functions:-

$$\begin{aligned} x'_c &= k_1(\theta_1, \ldots, \theta_6) \\ y'_c &= k_2(\theta_1, \ldots, \theta_6) \\ z'_c &= k_3(\theta_1, \ldots, \theta_6) \end{aligned}$$

[†] Some authors like to regard the forward kinematics as a co-ordinate transformation; but this is not possible since the spaces concerned are topologically different.

As we have seen previously, these are highly non-linear functions of the joint angles. However, if we are only interested in the neighbourhood of some point, it is possible to linearize the map. That is, we find a linear approximation to the original map. So if we make small changes in the joint angles we get:-

$$\delta x = \frac{\partial k_1}{\partial \theta_1} \delta\theta_1 + \frac{\partial k_1}{\partial \theta_2} \delta\theta_2 + \cdots + \frac{\partial k_1}{\partial \theta_6} \delta\theta_6$$

$$\delta y = \frac{\partial k_2}{\partial \theta_1} \delta\theta_1 + \frac{\partial k_2}{\partial \theta_2} \delta\theta_2 + \cdots + \frac{\partial k_2}{\partial \theta_6} \delta\theta_6$$

$$\delta z = \frac{\partial k_3}{\partial \theta_1} \delta\theta_1 + \frac{\partial k_3}{\partial \theta_2} \delta\theta_2 + \cdots + \frac{\partial k_3}{\partial \theta_6} \delta\theta_6$$

If we write $(\delta x_c', \delta y_c', \delta z_c')^T = \Delta \mathbf{x}$ and $(\delta\theta_1, \ldots, \delta\theta_6)^T = \Delta \boldsymbol{\theta}$, then we can summarize the above equations as:-

$$\Delta \mathbf{x} = \mathbf{J} \Delta \boldsymbol{\theta}$$

The matrix $\mathbf{J}$ is called the **jacobian** of the map; that is, the jacobian is the matrix of partial derivatives. In this case:-

$$\mathbf{J} = \begin{pmatrix} \frac{\partial k_1}{\partial \theta_1} & \frac{\partial k_1}{\partial \theta_2} & \cdots & \frac{\partial k_1}{\partial \theta_6} \\ \frac{\partial k_2}{\partial \theta_1} & \frac{\partial k_2}{\partial \theta_2} & \cdots & \frac{\partial k_2}{\partial \theta_6} \\ \frac{\partial k_3}{\partial \theta_1} & \frac{\partial k_3}{\partial \theta_2} & \cdots & \frac{\partial k_3}{\partial \theta_6} \end{pmatrix}$$

The jacobian matrix behaves very like the first derivative of a function of one variable. For a function of several variables we have a version of Taylor's theorem:-

$$\mathbf{x} + \Delta \mathbf{x} \approx \mathbf{k}(\boldsymbol{\theta}) + \mathbf{J}(\boldsymbol{\theta}) \Delta \boldsymbol{\theta}$$

For small variations about $\boldsymbol{\theta}$ the map is approximated by its value at $\boldsymbol{\theta}$ plus $\mathbf{J}(\boldsymbol{\theta})$ times the variation, $\Delta \boldsymbol{\theta}$.

For an example we turn to the planar manipulator yet again, see fig. 6.1. The kinematic equations of the end point are given by:-

$$x = l_1 \cos\theta_1 + l_2 \cos(\theta_1 + \theta_2) + l_3 \cos(\theta_1 + \theta_2 + \theta_3)$$

$$y = l_1 \sin\theta_1 + l_2 \sin(\theta_1 + \theta_2) + l_3 \sin(\theta_1 + \theta_2 + \theta_3)$$

The jacobian of this is:-

$$\mathbf{J}(\theta_1, \theta_2, \theta_3) = \begin{pmatrix} \frac{\partial x}{\partial \theta_1} & \frac{\partial x}{\partial \theta_2} & \frac{\partial x}{\partial \theta_3} \\ \frac{\partial y}{\partial \theta_1} & \frac{\partial y}{\partial \theta_2} & \frac{\partial y}{\partial \theta_3} \end{pmatrix}$$

where:-

$$\frac{\partial x}{\partial \theta_1} = -l_1 \sin\theta_1 - l_2 \sin(\theta_1 + \theta_2) - l_3 \sin(\theta_1 + \theta_2 + \theta_3)$$

$$\frac{\partial x}{\partial \theta_2} = l_2 \sin(\theta_1 + \theta_2) - l_3 \sin(\theta_1 + \theta_2 + \theta_3)$$

$$\frac{\partial x}{\partial \theta_3} = l_3 \sin(\theta_1 + \theta_2 + \theta_3)$$

$$\frac{\partial y}{\partial \theta_1} = l_1 \cos\theta_1 + l_2 \cos(\theta_1 + \theta_2) + l_3 \cos(\theta_1 + \theta_2 + \theta_3)$$

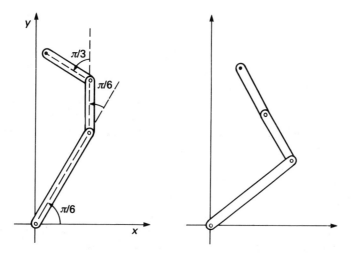

Figure 6.1 The Final and Starting Positions for the Planar Manipulator Example

$$\frac{\partial y}{\partial \theta_2} = l_2 \cos(\theta_1 + \theta_2) + l_3 \cos(\theta_1 + \theta_2 + \theta_3)$$
$$\frac{\partial y}{\partial \theta_3} = l_3 \cos(\theta_1 + \theta_2 + \theta_3)$$

6.2   Errors

One of the first uses we can make of the jacobian is to find the effect of errors in the joint angles. An error of $\mathbf{\Delta\theta}$ in the joint angles will produce a positional error of $\mathbf{\Delta x} = \mathbf{J}\,\mathbf{\Delta\theta}$. Because the map is non-linear, the effect of errors will be different at different positions. Consider the planar manipulator in its home position; $\theta_1 = \theta_2 = \theta_3 = 0$.

$$\mathbf{J}\,(0,0,0) = \begin{pmatrix} 0 & 0 & 0 \\ l_1 + l_2 + l_3 & l_2 + l_3 & l_3 \end{pmatrix}$$

To first order, no joint error can produce an error in the x-direction. An error of $1/10$ of a radian in θ_2, $\mathbf{\Delta\theta} = (0, 0.1, 0)^T$, will give a y-error of:-

$$\delta y \approx 0.1(l_2 + l_3)$$

In a different position, say $\theta_1 = \theta_3 = 0, \theta_2 = \frac{\pi}{2}$, the jacobian is:-

$$\mathbf{J}\,(0, \frac{\pi}{2}, 0) = \begin{pmatrix} -l_2 - l_3 & -l_2 - l_3 & -l_3 \\ l_1 & 0 & 0 \end{pmatrix}$$

Now an error $\mathbf{\Delta\theta} = (0, 0.1, 0)^T$ will give a positional error of:-

$$\mathbf{\Delta x} \approx \mathbf{J}\,(0, \frac{\pi}{2}, 0)\mathbf{\Delta\theta} = \begin{pmatrix} -\frac{1}{10}(l_2 + l_3) \\ 0 \end{pmatrix}$$

So there is no error in the y-direction and an error of $-\frac{1}{10}(l_2 + l_3)$ in the x-direction; to first order.

This tells us about singularities in the kinematics. In section 5.2 we defined a singularity as a point where the robot loses a degree-of-freedom. In fact at a singularity the robot loses an 'instantaneous' degree-of-freedom also. This means that, to first order, the robot's end-effector cannot move in one direction. The columns of the jacobian span the instantaneous directions the end-effector can move in. That is, the robot can only move in directions which are linear combinations of the columns of the jacobian. Thus a better definition of a singularity is as follows. A point ϕ in the joint space of a robot is a singular point if and only if the jacobian $\mathbf{J}(\phi)$ has less than maximal rank. That is, if there are linear dependencies among the columns of the jacobian.

In the example above, $\mathbf{J}(0,0,0)$ had a row of zeroes. So all the 2×2 submatrices would have zero determinant and thus the rank of the jacobian is one. Hence, the home position is singular. However, $\mathbf{J}(0, \frac{\pi}{2}, 0)$ has a submatrix with non-zero determinant, so the rank is two, which is the maximum and the point is thus non-singular. If we are interested in the position and orientation of a six joint manipulator then the jacobian is a square matrix. In such cases the condition for a point ϕ to be singular reduces to $\det(\mathbf{J}(\phi)) = 0$; that is, the matrix is singular.

6.3 Numerical Methods

The jacobian of a manipulator also finds applications in various numerical methods, for example, to solve the inverse kinematics. As an example, we will look at a method which is the many-variable extension of the Newton-Raphson method.

For a single variable the Newton-Raphson method is as follows. We wish to solve an equation:-

$$f(x) = 0$$

for some function f. We begin with an initial guess $x^{(0)}$, and then refine this guess using the iteration formula:-

$$x^{(i+1)} = x^{(i)} - \frac{f(x^{(i)})}{\frac{\mathrm{d}}{\mathrm{d}x}f(x^{(i)})}$$

Here the notation, superscript (i), denotes the ith iterate.

This generalizes to many variables quite easily. Suppose we have six equations to solve in six variables:-

$$
\begin{aligned}
f_1(\theta_1, \ldots, \theta_6) &= 0 \\
f_2(\theta_1, \ldots, \theta_6) &= 0 \\
&\vdots \\
f_6(\theta_1, \ldots, \theta_6) &= 0
\end{aligned}
$$

We may summarize this with the vector notation as $\mathbf{f}(\boldsymbol{\theta}) = \mathbf{0}$. Taylor's theorem tells us that:-

$$\mathbf{f}(\boldsymbol{\theta} + \mathbf{h}) \approx \mathbf{f}(\boldsymbol{\theta}) + \mathbf{J}(\boldsymbol{\theta})\mathbf{h}$$

Now if we assume that $\boldsymbol{\theta}$ is the root we are looking for, then since $\mathbf{f}(\boldsymbol{\theta}) = \mathbf{0}$, we can approximate the error $\mathbf{h}$ as:-

$$\mathbf{h} \approx \mathbf{J}^{-1}(\boldsymbol{\theta})\mathbf{f}(\boldsymbol{\theta} + \mathbf{h})$$

Since at this stage, we do not know $\boldsymbol{\theta}$ we cannot calculate $\mathbf{J}^{-1}(\boldsymbol{\theta})$, so we approximate it by $\mathbf{J}^{-1}(\boldsymbol{\theta} + \mathbf{h})$ which is our guess. By setting $\mathbf{h}^{(i)} = \boldsymbol{\theta}^{(i)} - \boldsymbol{\theta}^{(i+1)}$ we can set up the following iterative scheme:-

$$\boldsymbol{\theta}^{(i+1)} = \boldsymbol{\theta}^{(i)} - \mathbf{J}^{-1}(\boldsymbol{\theta}^{(i)})\mathbf{f}(\boldsymbol{\theta}^{(i)})$$

This is the Newton-Raphson formula for many variables. In practice, however, inverting matrices is very slow. A quicker method is to solve the linear equations $\mathbf{J}(\boldsymbol{\theta}^{(i)})\mathbf{h}^{(i)} = \mathbf{f}(\boldsymbol{\theta}^{(i)})$, using Gauss elimination, for example.

To see how this could be used, we look at a simple example. Consider the planar manipulator once more. This time we want to take account of the output angle $\Phi = \theta_1 + \theta_2 + \theta_3$ as well as the position of the end point. We will assume that $l_1 = 2, l_2 = 1$ and $l_3 = 1$ in some units. Suppose the arm is in the position illustrated in fig. 6.1, where $\theta_1 = \pi/3, \theta_2 = \pi/6$ and $\theta_3 = \pi/6$.

The forward kinematics gives the starting position as:-

$$
\begin{aligned}
x &= 2\cos(\frac{\pi}{3}) + \cos(\frac{\pi}{3} + \frac{\pi}{6}) + \cos(\frac{\pi}{3} + \frac{\pi}{6} + \frac{\pi}{6}) &= 0.5000 \\
y &= 2\sin(\frac{\pi}{3}) + \sin(\frac{\pi}{3} + \frac{\pi}{6}) + \sin(\frac{\pi}{3} + \frac{\pi}{6} + \frac{\pi}{6}) &= 3.5981 \\
\Phi &= \frac{\pi}{3} + \frac{\pi}{6} + \frac{\pi}{6} &= 2\frac{\pi}{3}
\end{aligned}
$$

Now suppose we want to move the end-effector to the position where:-

$$x = 0.5, \qquad y = 3.0, \qquad \Phi = \frac{2\pi}{3}$$

We set up the three functions:-

$$
\begin{aligned}
f_1 &= 2\cos(\theta_1) + \cos(\theta_1 + \theta_2) + \cos(\theta_1 + \theta_2 + \theta_3) - 0.5 \\
f_2 &= 2\sin(\theta_1) + \sin(\theta_1 + \theta_2) + \sin(\theta_1 + \theta_2 + \theta_3) - 3.0 \\
f_3 &= \theta_1 + \theta_2 + \theta_3 - \frac{2\pi}{3}
\end{aligned}
$$

In the desired position all three of these functions will vanish. So we may use the Newton-Raphson method to find the roots, that is the values of the joint angles. The jacobian has columns:-

$$\mathbf{J}(\theta_1, \theta_2, \theta_3) =$$
$$\begin{pmatrix} -2\sin\theta_1 - \sin(\theta_1 + \theta_2) & -\sin(\theta_1 + \theta_2) & \sin(\theta_1 + \theta_2 + \theta_3) \\ -\sin(\theta_1 + \theta_2 + \theta_3) & -\sin(\theta_1 + \theta_2 + \theta_3) & \\ 2\cos\theta_1 + \cos(\theta_1 + \theta_2) & \cos(\theta_1 + \theta_2) & \cos(\theta_1 + \theta_2 + \theta_3) \\ +\cos(\theta_1 + \theta_2 + \theta_3) & +\cos(\theta_1 + \theta_2 + \theta_3) & \\ 1 & 1 & 1 \end{pmatrix}$$

As our initial guess we may as well use the starting position, so that $\theta^{(0)} = (\pi/3, \pi/6, \pi/6)^T$. So now:-

$$\mathbf{J}(\theta^{(0)}) = \begin{pmatrix} -3.5981 & -1.8660 & -0.8660 \\ 0.5000 & -0.5000 & -0.5000 \\ 1 & 1 & 1 \end{pmatrix} \quad \text{and} \quad \mathbf{f}(\theta^{(0)}) = \begin{pmatrix} 0.0000 \\ 0.5981 \\ 0.0000 \end{pmatrix}$$

We find the first approximation to the error by solving:-

$$\begin{pmatrix} -3.5981 & -1.8660 & -0.8660 \\ 0.5000 & -0.5000 & -0.5000 \\ 1 & 1 & 1 \end{pmatrix} \begin{pmatrix} h_1^{(0)} \\ h_2^{(0)} \\ h_3^{(0)} \end{pmatrix} = \begin{pmatrix} 0.0000 \\ 0.5981 \\ 0.0000 \end{pmatrix}$$

To four decimal places the solution is:-

$$\begin{pmatrix} h_1^{(0)} \\ h_2^{(0)} \\ h_3^{(0)} \end{pmatrix} = \begin{pmatrix} 0.5981 \\ -1.6340 \\ 1.0359 \end{pmatrix} \quad \text{and thus} \quad \theta^{(1)} = \theta^{(0)} - h^{(0)} = \begin{pmatrix} 0.4491 \\ 2.1576 \\ -0.5123 \end{pmatrix}$$

For the next iteration the values of the jacobian and the functions are:-

$$\mathbf{J}(\theta^{(1)}) = \begin{pmatrix} 2.2441 & -1.3758 & -0.8660 \\ 0.4413 & -1.4603 & -0.5000 \\ 1 & 1 & 1 \end{pmatrix} \quad \text{and} \quad \mathbf{f}(\theta^{(1)}) = \begin{pmatrix} -0.059 \\ -0.756 \\ 0.0000 \end{pmatrix}$$

This then gives:-

$$h^{(1)} = \begin{pmatrix} -0.1606 \\ 0.5493 \\ -0.3887 \end{pmatrix} \quad \text{and therefore} \quad \theta^{(2)} = \begin{pmatrix} 0.6097 \\ 1.6083 \\ -0.1236 \end{pmatrix}$$

The values of the functions here are now $\mathbf{f}(\theta^{(2)}) = (0.0367, -0.1910, 0.0000)^T$, which is getting closer to zero. The next two iterations give:-

$$\theta^{(3)} = \begin{pmatrix} 0.6789 \\ 1.4106 \\ 0.0049 \end{pmatrix}, \quad \mathbf{f}(\theta^{(3)}) = \begin{pmatrix} 0.0608 \\ -0.0096 \\ 0.0000 \end{pmatrix}$$

and

$$\theta^{(4)} = \begin{pmatrix} 0.6984 \\ 1.4329 \\ -0.0369 \end{pmatrix}, \quad \mathbf{f}(\theta^{(4)}) = \begin{pmatrix} 0.0001 \\ -0.0010 \\ 0.0000 \end{pmatrix}$$

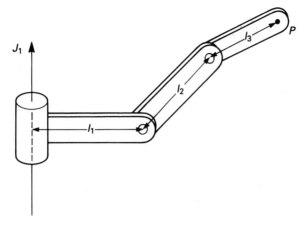

Figure 6.2 A Three Joint Manipulator

If we can tolerate an accuracy of only two decimal places we can stop here. Otherwise we could continue to any desired accuracy. In this particular case we have an exact solution of the inverse kinematics: compare the results here with those of exercise 5.3.

Exercises

6.1 A manipulator has the kinematic structure illustrated in fig. 6.2.

(i) By setting up a suitable co-ordinate system and home position, find the kinematic equations for the co-ordinates of the point P.

(ii) Calculate the jacobian of this manipulator.

(iii) Show that the limiting positions, where the determinant of the jacobian vanishes, lie on the surface of a hollow torus. Assume that $l_1 > l_2 + l_3$.

(iv) How many postures are there in general?

(v) If $l_1 < (l_2 + l_3)$, show that there exist points with four postures, and that points on the J_1 axis have a continuous set of postures.

6.2 For a parallel manipulator it is the inverse kinematics that gives a mapping, this time from the space of rigid body motions to joint space. Find the jacobian matrix for the parallel planar manipulator whose inverse kinematics were found in exercise 5.7.

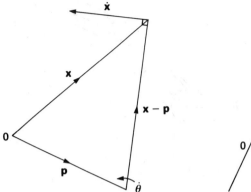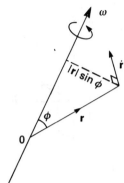

Figure 6.3 (a) Linear Velocity Given by an Instan-   (b) Relation Between Angular and Lin-
taneous Centre of Rotation   ear Velocities

6.4   Linear Velocities

Perhaps the most important use for jacobians is for relating the joint velocities to the link
velocities. In section 6.1 we saw that $\Delta x \approx \mathbf{J} \Delta \theta$. Dividing by Δt and proceeding to the
limit we obtain the exact relation:-

$$\dot{\mathbf{x}} = \mathbf{J} \dot{\theta}$$

The dots, as usual, denote differentiation with respect to time. This is quite general, but
usually we are interested in the linear velocity of some point on a link, or the angular velocity
of a link. The movements that can be performed by robots are very general; however, for
any rigid body motion in the plane there is always a centre of rotation. Similarly for motion
on the surface of a sphere, as one gets from a spherical wrist, there is always an axis of
rotation. For rigid movements in three dimensions there is always a fixed line; the screw
axis. If a rigid body undergoes some complicated motion in the plane, for example, then at
any time in the body's motion there will be an instantaneous centre of rotation. Similarly
we get instantaneous rotation axes and instantaneous screw axes. As we shall see below,
these concepts are closely related to the velocities that we are interested in.

In two dimensions we have a simple relation between the velocity of a point and the
instantaneous centre of rotation, see fig. 6.3(a). This can also be shown using the 3×3
matrices which represent rigid movements. Let $\mathbf{p}$ be the centre of rotation and suppose
that we wish to know the velocity of the point $\mathbf{x}$. Now the position of $\mathbf{x}$ is given by:-

$$\begin{pmatrix} \mathbf{x}(t) \\ 1 \end{pmatrix} = \begin{pmatrix} \mathbf{R}(\theta) & (\mathbf{I} - \mathbf{R}(\theta))\mathbf{p} \\ 0 & 1 \end{pmatrix} \begin{pmatrix} \mathbf{x}(0) \\ 1 \end{pmatrix}$$

Note that from now on we will not write the partition lines in partitional matrices. Now
assume that $\theta = 0$ when $t = 0$. We can always arrange for this to be true by beginning the
measurements from the point we are interested in. Now at $\theta = 0$ the time derivative of the

above is given by:-

$$\begin{pmatrix} \dot{\mathbf{x}}(0) \\ 0 \end{pmatrix} = \begin{pmatrix} \dot{\mathbf{R}}(0)\dot{\theta} & -\dot{\mathbf{R}}(0)\mathbf{p}\dot{\theta} \\ 0 & 0 \end{pmatrix} \begin{pmatrix} \mathbf{x}(0) \\ 1 \end{pmatrix}$$

This gives the equation:-

$$\dot{\mathbf{x}}(0) = \dot{\mathbf{R}}(0)(\mathbf{x}(0) - \mathbf{p})\dot{\theta}$$

But we know that $\mathbf{R} = \begin{pmatrix} \cos\theta & -\sin\theta \\ \sin\theta & \cos\theta \end{pmatrix}$, so taking the differential and setting $\theta = 0$ gives:-

$$\dot{\mathbf{R}}(0) = \begin{pmatrix} 0 & -1 \\ 1 & 0 \end{pmatrix}$$

As mentioned above, we may begin measuring time anywhere, so these results apply for any time, not just $t = 0$. We can drop the time dependence and write:-

$$\dot{x} = (p_y - y)\dot{\theta}$$
$$\dot{y} = (x - p_x)\dot{\theta}$$

Notice that the vector $\dot{\mathbf{x}}$ is always normal to $(\mathbf{x} - \mathbf{p})$. These results can be used to compute the jacobian of planar manipulators. For a three joint planar manipulator we have:-

$$\begin{pmatrix} \mathbf{x}(t) \\ 1 \end{pmatrix} = \mathbf{A}_1(\theta_1)\mathbf{A}_2(\theta_2)\mathbf{A}_3(\theta_3) \begin{pmatrix} \mathbf{x}(0) \\ 1 \end{pmatrix}$$

Again we can arrange things so that at the point of interest $\theta_1 = \theta_2 = \theta_3 = 0$. Then since $\mathbf{A}_i(0)$ is the identity matrix, when we differentiate and set the joint angles to zero, we get:-

$$\begin{pmatrix} \dot{\mathbf{x}}(0) \\ 0 \end{pmatrix} = \dot{\mathbf{A}}_1(0) \begin{pmatrix} \mathbf{x}(0) \\ 1 \end{pmatrix} \dot{\theta}_1 + \dot{\mathbf{A}}_2(0) \begin{pmatrix} \mathbf{x}(0) \\ 1 \end{pmatrix} \dot{\theta}_2 + \dot{\mathbf{A}}_3(0) \begin{pmatrix} \mathbf{x}(0) \\ 1 \end{pmatrix} \dot{\theta}_3$$

Once again there is nothing special about the point $\theta_1 = \theta_2 = \theta_3 = 0$, and our result applies quite generally. For each $\mathbf{A}$-matrix the centre of rotation is simply the current position of the joint. So if we denote the current position of joint i by $\mathbf{j}_i$ we have:-

$$\dot{x} = (j_{1y} - y)\dot{\theta}_1 + (j_{2y} - y)\dot{\theta}_2 + (j_{3y} - y)\dot{\theta}_3$$
$$\dot{y} = (x - j_{1x})\dot{\theta}_1 + (x - j_{2x})\dot{\theta}_2 + (x - j_{3x})\dot{\theta}_3$$

This can be neatly summarized as:-

$$\begin{pmatrix} \dot{x} \\ \dot{y} \end{pmatrix} = \begin{pmatrix} (j_{1y} - y) & (j_{2y} - y) & (j_{3y} - y) \\ (x - j_{1x}) & (x - j_{2x}) & (x - j_{3x}) \end{pmatrix} \begin{pmatrix} \dot{\theta}_1 \\ \dot{\theta}_2 \\ \dot{\theta}_3 \end{pmatrix}$$

And this shows us that the jacobian is given by:-

$$\mathbf{J}(\theta_1, \theta_2, \theta_3) = \begin{pmatrix} (j_{1y} - y) & (j_{2y} - y) & (j_{3y} - y) \\ (x - j_{1x}) & (x - j_{2x}) & (x - j_{3x}) \end{pmatrix}$$

The suitably altered co-ordinates of the joints are the columns of the jacobian.

The planar manipulator in section 6.1 had its joints at:-

$$\begin{pmatrix} j_{1x} \\ j_{1y} \end{pmatrix} = \begin{pmatrix} 0 \\ 0 \end{pmatrix}, \quad \begin{pmatrix} j_{2x} \\ j_{2y} \end{pmatrix} = \begin{pmatrix} l_1 \cos\theta_1 \\ l_1 \sin\theta_1 \end{pmatrix},$$

$$\begin{pmatrix} j_{3x} \\ j_{3y} \end{pmatrix} = \begin{pmatrix} l_1 \cos\theta_1 + l_2 \cos(\theta_1 + \theta_2) \\ l_1 \sin\theta_1 + l_2 \sin(\theta_1 + \theta_2) \end{pmatrix}$$

and the end point has co-ordinates:-

$$x = l_1 \cos\theta_1 + l_2 \cos(\theta_1 + \theta_2) + l_3 \cos(\theta_1 + \theta_2 + \theta_3)$$
$$y = l_1 \sin\theta_1 + l_2 \sin(\theta_1 + \theta_2) + l_3 \sin(\theta_1 + \theta_2 + \theta_3)$$

So the jacobian is exactly as calculated in section 6.1. Its columns are:-

$$\begin{pmatrix} \frac{\partial x}{\partial \theta_1} \\ \frac{\partial y}{\partial \theta_1} \end{pmatrix} = \begin{pmatrix} -l_1 \sin\theta_1 - l_2 \sin(\theta_1 + \theta_2) - l_3 \sin(\theta_1 + \theta_2 + \theta_3) \\ l_1 \cos\theta_1 + l_2 \cos(\theta_1 + \theta_2) + l_3 \cos(\theta_1 + \theta_2 + \theta_3) \end{pmatrix}$$

$$\begin{pmatrix} \frac{\partial x}{\partial \theta_2} \\ \frac{\partial y}{\partial \theta_2} \end{pmatrix} = \begin{pmatrix} l_2 \sin(\theta_1 + \theta_2) - l_3 \sin(\theta_1 + \theta_2 + \theta_3) \\ l_2 \cos(\theta_1 + \theta_2) + l_3 \cos(\theta_1 + \theta_2 + \theta_3) \end{pmatrix}$$

$$\begin{pmatrix} \frac{\partial x}{\partial \theta_3} \\ \frac{\partial y}{\partial \theta_3} \end{pmatrix} = \begin{pmatrix} l_3 \sin(\theta_1 + \theta_2 + \theta_3) \\ l_3 \cos(\theta_1 + \theta_2 + \theta_3) \end{pmatrix}$$

but here we have not had to find any derivatives.

6.5 Angular Velocities

The angular velocity of a rigid body is a vector. It is aligned along the instantaneous rotation axis of the body and its magnitude is the angular speed about the axis. Consider a point $\mathbf{r}$ attached to a body rotating with angular velocity ω; see fig. 6.3(b). The linear velocity of the point is given by $\dot{\mathbf{r}} = \omega \wedge \mathbf{r}$. If we represent the rotations by 3×3 matrices we have that:-

$$\mathbf{r}(t) = \mathbf{R}(t)\mathbf{r}(0)$$

Differentiating and setting $t = 0$ gives:-

$$\dot{\mathbf{r}}(0) = \dot{\mathbf{R}}(0)\mathbf{r}(0)$$

Comparing this with our first result shows that $\dot{\mathbf{R}}(0)$ must have the same effect on vectors as '$\omega \wedge$'; in other words for any vector $\mathbf{a}$ we must have:-

$$\dot{\mathbf{R}}(0)\mathbf{a} = \omega \wedge \mathbf{a}$$

This is not hard to solve, see exercise 2.7. It gives us that:-

$$\dot{\mathbf{R}}(0) = \begin{pmatrix} 0 & -\omega_z & \omega_y \\ \omega_z & 0 & -\omega_x \\ -\omega_y & \omega_x & 0 \end{pmatrix}$$

This can be used to find the velocity of the last link of a spherical wrist. For a three joint wrist the overall transformation $\mathbf{K}$ is the product of three rotations, $\mathbf{K} = \mathbf{R}_1(\theta_1)\mathbf{R}_2(\theta_2)\mathbf{R}_3(\theta_3)$. The derivative when all the joint angles are zero is:-

$$\dot{\mathbf{K}} = \dot{\mathbf{R}}_1 + \dot{\mathbf{R}}_2 + \dot{\mathbf{R}}_3$$

and hence the angular velocity of the final link is just the sum of the angular velocities of the joints. Now, because each joint just rotates about its axis, we can write the angular velocities of the joints as $\boldsymbol{\omega}_i = \hat{\mathbf{v}}_i \dot{\theta}_i$; where $\hat{\mathbf{v}}_i$ is the unit vector along the ith joint. The angular velocity of the last link can be expressed by the following matrix equation:-

$$\boldsymbol{\omega} = \begin{pmatrix} \hat{v}_{1x} & \hat{v}_{2x} & \hat{v}_{3x} \\ \hat{v}_{1y} & \hat{v}_{2y} & \hat{v}_{3y} \\ \hat{v}_{1z} & \hat{v}_{2z} & \hat{v}_{3z} \end{pmatrix} \begin{pmatrix} \dot{\theta}_1 \\ \dot{\theta}_2 \\ \dot{\theta}_3 \end{pmatrix}$$

Notice that the columns of the jacobian here are just the vectors along the joint axes. So it is easy now to calculate the jacobian of the 3-R wrist, introduced in section 4.2, for example.

$$\hat{\mathbf{v}}_1 = \begin{pmatrix} 0 \\ 0 \\ 1 \end{pmatrix}, \quad \hat{\mathbf{v}}_2 = \mathbf{R}(\theta_1, \mathbf{k}) \begin{pmatrix} 0 \\ 1 \\ 0 \end{pmatrix} = \begin{pmatrix} -\sin\theta_1 \\ \cos\theta_1 \\ 0 \end{pmatrix},$$

$$\hat{\mathbf{v}}_3 = \mathbf{R}(\theta_1, \mathbf{k})\mathbf{R}(\theta_2, \mathbf{j}) \begin{pmatrix} 0 \\ 0 \\ 1 \end{pmatrix} = \begin{pmatrix} \cos\theta_1 \sin\theta_2 \\ \sin\theta_1 \sin\theta_2 \\ \cos\theta_2 \end{pmatrix}$$

Hence the jacobian for this manipulator is:-

$$\mathbf{J} = \begin{pmatrix} 0 & -\sin\theta_1 & \cos\theta_1 \sin\theta_2 \\ 0 & \cos\theta_1 & \sin\theta_1 \sin\theta_2 \\ 1 & 0 & \cos\theta_2 \end{pmatrix}$$

6.6   Combining Linear and Angular Velocities

For a rigid body moving in three dimensions we want to know both its angular velocity and the linear velocity of its points. We can find these by considering a general screw motion:-

$$\begin{pmatrix} \mathbf{R} & \mathbf{t} \\ 0 & 1 \end{pmatrix}$$

In section 2.6 we saw that the translation vector is given by $\mathbf{t} = \frac{\theta p}{2\pi}\hat{\mathbf{v}} + (\mathbf{I} - \mathbf{R})\mathbf{u}$, where p is the pitch of the screw, $\hat{\mathbf{v}}$ a unit vector along its axis, $\mathbf{u}$ a point on the axis and we have used θ, rather than ϕ, for the joint variable which depends on time. The velocity of a point $\mathbf{x}$ is then given by:-

$$\begin{pmatrix} \dot{\mathbf{x}} \\ 0 \end{pmatrix} = \begin{pmatrix} \dot{\mathbf{R}} & \dot{\mathbf{t}} \\ 0 & 0 \end{pmatrix} \begin{pmatrix} \mathbf{x} \\ 1 \end{pmatrix}$$

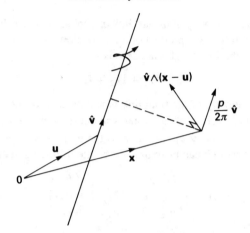

Figure 6.4 A Screw Motion

Using what we already know about the derivatives of rotation matrices, this can be written as:-

$$\dot{\mathbf{x}} = \boldsymbol{\omega} \wedge \mathbf{x} + \mathbf{s}$$

The linear velocity $\mathbf{s} = \dot{\mathbf{t}}$ is a characteristic velocity of the motion. Physically it is the linear velocity of points on a line through the origin parallel to the axis of rotation. We can find $\mathbf{s}$ by differentiating $\mathbf{t}$:-

$$\mathbf{s} = \dot{\mathbf{t}} = \dot{\theta}\frac{p}{2\pi}\hat{\mathbf{v}} - \dot{\mathbf{R}}\mathbf{u} = \dot{\theta}\frac{p}{2\pi}\hat{\mathbf{v}} - \boldsymbol{\omega} \wedge \mathbf{u}$$

Notice that the term $\dot{\theta}\frac{p}{2\pi}\hat{\mathbf{v}}$ is the velocity of a point lying on the screw axis. So, as we would expect from contemplating fig. 6.4, the velocity of a point $\mathbf{x}$ can be written more fully as:-

$$\dot{\mathbf{x}} = \boldsymbol{\omega} \wedge (\mathbf{x} - \mathbf{u}) + \dot{\theta}\frac{p}{2\pi}\hat{\mathbf{v}}$$

We can combine the angular and linear velocities into six component vectors $\begin{pmatrix} \boldsymbol{\omega} \\ \mathbf{s} \end{pmatrix}$. These six component vectors are called **instantaneous screws**. As we shall see, they are for rigid bodies the analogue of the angular velocity of particles.

Now, if the screw motion is about a joint, then $\hat{\mathbf{v}}$ is the unit vector in the direction of the joint, $\mathbf{u}$ is the position vector of a point on the joint axis and the angular velocity will be $\boldsymbol{\omega} = \hat{\mathbf{v}}\dot{\theta}$. Finally p is the pitch of the joint. The velocity of a point attached to the joint will be given by:-

$$\dot{\mathbf{x}} = (\hat{\mathbf{v}} \wedge (\mathbf{x} - \mathbf{u}) + \frac{p}{2\pi}\hat{\mathbf{v}})\dot{\theta}$$

Connecting six joints together, as in a serial robot, both the linear and angular velocities add vectorially to give the angular velocity $\boldsymbol{\omega}$, and the linear velocity $\mathbf{s}$, of the last link. So we may write this in terms of instantaneous screws:-

$$\begin{pmatrix} \omega \\ s \end{pmatrix} = \begin{pmatrix} \hat{\mathbf{v}}_1 \\ \mathbf{u}_1 \wedge \hat{\mathbf{v}}_1 + \frac{p_1}{2\pi}\hat{\mathbf{v}}_1 \end{pmatrix} \dot{\theta}_1 + \begin{pmatrix} \hat{\mathbf{v}}_2 \\ \mathbf{u}_2 \wedge \hat{\mathbf{v}}_2 + \frac{p_2}{2\pi}\hat{\mathbf{v}}_2 \end{pmatrix} \dot{\theta}_2 + \cdots + \begin{pmatrix} \hat{\mathbf{v}}_6 \\ \mathbf{u}_6 \wedge \hat{\mathbf{v}}_6 + \frac{p_1}{2\pi}\hat{\mathbf{v}}_6 \end{pmatrix} \dot{\theta}_6$$

This can be condensed into the matrix equation:-

$$\begin{pmatrix} \omega \\ s \end{pmatrix} = \mathbf{J} \begin{pmatrix} \dot{\theta}_1 \\ \dot{\theta}_2 \\ \dot{\theta}_3 \\ \dot{\theta}_4 \\ \dot{\theta}_5 \\ \dot{\theta}_6 \end{pmatrix}$$

The columns of the jacobian matrix $\mathbf{J}$ are given by $\begin{pmatrix} \hat{\mathbf{v}}_i \\ \mathbf{u}_i \wedge \hat{\mathbf{v}}_i + \frac{p_i}{2\pi}\hat{\mathbf{v}}_i \end{pmatrix}$, and are determined by the ith joint. In other words, to each joint there is an associated instantaneous screw which depends only on the position, orientation and pitch of the joint. These 'joint screws' are reasonably easy to calculate and once again we have been able to find the jacobian matrix of the manipulator without computing any partial derivatives.

The jacobian is of fundamental importance in robotics. This is because it is the linearization of the forward kinematics. Hence, it tells us about errors, velocities and other first order properties of the robot. For robot manipulators with an open loop structure the jacobian is very simple to calculate since its columns are given by the joints of the robot.

Exercises

6.3 In two dimensions a sliding joint is represented by a matrix $\begin{pmatrix} \mathbf{I} & \mathbf{x} \\ 0 & 1 \end{pmatrix}$, where $\mathbf{x} = \begin{pmatrix} \alpha \\ \beta \end{pmatrix} d$, with $d = d(t)$ is a function of time.

    (i) Find the velocity of a point undergoing such a translation. If such a joint were used in a serial manipulator, what would be the corresponding column in the jacobian?

    (ii) A planar manipulator consists of a revolute joint and a sliding joint. In the home configuration the revolute joint is at the origin, the sliding joint is aligned along the x-axis and a point Q attached to the last link has co-ordinates $(1, 0)$. Find the velocity of Q as a function of the joint angle θ of the revolute joint and the extension d of the sliding joint.

6.4 Find the jacobian of the Cincinnati T^3 wrist illustrated in fig. 4.4.

6.5 Let $\mathbf{R}(t)$ be a one parameter family of rotation matrices, such that $\mathbf{R}(0) = \mathbf{I}$. Using the fact that rotation matrices satisfy $\mathbf{R}\mathbf{R}^T = \mathbf{I}$, show that $\dot{\mathbf{R}}(0)$ is an antisymmetric matrix.

6.6   The instantaneous screw of a revolute joint is given by $\begin{pmatrix} \hat{\mathbf{v}} \\ \mathbf{u} \wedge \hat{\mathbf{v}} \end{pmatrix}$. If a rotation $\mathbf{R}$ followed by a translation $\mathbf{t}$ is performed on the joint, show that the new joint will have the instantaneous screw given by:-

$$\begin{pmatrix} \mathbf{R} & 0 \\ \mathbf{TR} & \mathbf{R} \end{pmatrix} \begin{pmatrix} \hat{\mathbf{v}} \\ \mathbf{u} \wedge \hat{\mathbf{v}} \end{pmatrix}$$

where $\mathbf{T} = \begin{pmatrix} 0 & -t_z & t_y \\ t_z & 0 & -t_x \\ -t_y & t_x & 0 \end{pmatrix}$.

7 Trajectory Following

7.1 Following Paths

So far we have only considered moving a robot arm to a particular point. In many applications, however, we want the robot to follow a prescribed path. Examples might be seam welding, profile and pattern cutting, or applying adhesives.

Here a path is a continuous sequence of rigid body motions, $\mathbf{K}(t)$; the parameter t represents time. So we must solve the following equation:-

$$\mathbf{A}_1(\theta_1)\mathbf{A}_2(\theta_2)\mathbf{A}_3(\theta_3)\mathbf{A}_4(\theta_4)\mathbf{A}_5(\theta_5)\mathbf{A}_6(\theta_6) = \mathbf{K}(t)$$

that is, we must find the joint angles as functions of time, $\theta_i(t)$. This is a formidable task: in some special cases it may be possible to find the exact solution, but generally it is impossible. In practice we approximate these functions, $\theta_i(t)$. This is a good idea in any case since a computer will be used to calculate the functions and computers can only compute approximations to the desired functions.

Usually, since a computer is controlling the robot, we will need to know the values of the functions every fraction of a second. The precise time interval will depend on how quickly the computer can respond. For very accurate work it is necessary to compute the inverse kinematics at each of these points. This is accurate if the inverse kinematics are known, since it involves no approximation (except for the usual approximations involved in computing inverse trigonometric functions and square roots by computer). However, it is very slow, because the inverse kinematics is so computationally intensive. Typically a floating point multiplication will take something like 10^{-6} seconds on a modern microcomputer, a large mainframe computer may do this a hundred times quicker; but on either, computing the inverse sine of an angle will take ten times as long. Since these calculations must be performed every millimetre or so along the path, it is often the time taken to calculate the inverse kinematics which determines the speed of the robot.

It is possible to speed up these calculations in several ways. For example, rather than recalculate inverse trigonometric functions every time, it is possible to store the values in the form of a look-up table. Intermediate values can be approximated by some interpolation scheme. This method saves time at the expense of memory; the values must be kept in semiconductor memory or a gain in speed will not be achieved. One could go even further, and store the values of all six inverse kinematic functions; however, this will usually take far too much memory. In the end, the success of these methods depends on a subtle mixture of mass storage and efficient interpolating algorithms.

Another approach is to approximate the desired functions directly. The inverse kinematics can be used for some values of t, and then any method of interpolation can be used to approximate the values of the functions at intermediate times. In this way the problem is converted to a straightforward problem in numerical analysis. It is these methods that we study in the rest of this chapter.

7.2   Linear Approximations

The paths we want to follow may be straight lines or arcs of circles in the simplest cases. But there is no reason to restrict ourselves to such paths. In general, however, the path taken by the robot will consist of three sections. To begin with the robot is at rest and in the first path segment we must accelerate it. This is sometimes called the lift-off phase of the trajectory. During the second path segment we expect the robot to be moving with constant speed. Finally we must decelerate the robot to rest. This is referred to as the set-down phase. However, we begin by considering the simplest case of uniform motion along a straight line.

Let us look at moving the wrist centre of the Puma. Recall exercise 5.5. There we assumed the dimensions of the Puma to be $L_2 = 4$, $D_3 = 1$ and $D_4 = 4$, in some system of units. In the exercise we used the inverse kinematics to find the joint angles required to place the wrist centre at the two points:-

$$\mathbf{p}_c = \begin{pmatrix} 5/\sqrt{2} \\ -3/\sqrt{2} \\ -4 \end{pmatrix} \quad \text{and} \quad \mathbf{p}_c = \begin{pmatrix} -(2 + \frac{3}{\sqrt{2}}) \\ -(2 + \frac{5}{\sqrt{2}}) \\ \frac{4}{\sqrt{2}} \end{pmatrix}$$

Suppose we want to move the wrist centre along a straight line between these two points. We can normalize the units of time so that at time $t = 0$ the wrist is at the first point and then at time $t = 1$ the wrist must reach the second point. If the wrist moves with uniform velocity along the line joining the two points, then at any intermediate time the co-ordinates of the wrist centre will be:-

$$\mathbf{p}_c(t) = (1 - t) \begin{pmatrix} 5/\sqrt{2} \\ -3/\sqrt{2} \\ -4 \end{pmatrix} + t \begin{pmatrix} -(2 + \frac{3}{\sqrt{2}}) \\ -(2 + \frac{5}{\sqrt{2}}) \\ \frac{4}{\sqrt{2}} \end{pmatrix}$$

Now, from the inverse kinematics we know the possible values of the joint angles which place the wrist at the start and finish positions. However, this is really too much information. Which of the possible solutions do we use? Usually the joint angles at the beginning of the

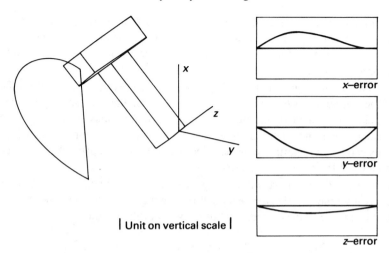

Figure 7.1 Linear Approximation in Joint Variables to Straight Line Path in Cartesian Variables

motion are given, since the history of how the arm reached the start position will be known. In general we will have several possible sets of joint angles that achieve the finishing position. Any of these could be chosen, but normally it is wise to keep to the same posture as at the start position. This is because on most robots it is not possible to change posture without encountering a singularity. Usually it is hard to control the robot in the vicinity of a singularity, thus singularities are to be avoided.

We will choose the following solutions for the start and finish positions:-

$$\boldsymbol{\theta}(0) = \begin{pmatrix} \theta_1(0) \\ \theta_2(0) \\ \theta_3(0) \end{pmatrix} = \begin{pmatrix} \pi/4 \\ \pi/2 \\ \pi/2 \end{pmatrix}, \qquad \boldsymbol{\theta}(1) = \begin{pmatrix} \theta_1(1) \\ \theta_2(1) \\ \theta_3(1) \end{pmatrix} = \begin{pmatrix} -\pi/4 \\ \pi/4 \\ \pi/4 \end{pmatrix}$$

The simplest approximation to use would be a linear one. Let us denote the approximation by $\boldsymbol{\theta}_l(t)$. It will be given by:-

$$\boldsymbol{\theta}_l(t) = (1-t)\boldsymbol{\theta}(0) + t\boldsymbol{\theta}(1) = (1-t) \begin{pmatrix} \pi/4 \\ \pi/2 \\ \pi/2 \end{pmatrix} + t \begin{pmatrix} -\pi/4 \\ \pi/4 \\ \pi/4 \end{pmatrix}$$

Notice, that when $t = 0$ this is just $\boldsymbol{\theta}(0)$, while at $t = 1$ the approximation reduces to $\boldsymbol{\theta}(1)$. These are the values we found from the inverse kinematics, so at least at $t = 0$ and $t = 1$ our approximation will be correct.

We can use the forward kinematics to calculate the path of the wrist centre that this approximation would generate. The difference between this and the desired straight line path is plotted in fig. 7.1. Not too surprisingly, near the end points we have good agreement, signified by the difference being near zero. However, around the middle of the path the error is large.

One way to get a more accurate approximation is to use several short path segments rather than one long one. We divide the desired straight line path into several short segments and use a different linear approximation for each segment. This means that we will have to use the inverse kinematics for the end points of every segment. Again, there is a trade off between speed and accuracy.

We may choose the segments to have any length. However, as we shall see later, the choice of the end points of the segments will affect the accuracy of our approximation. In robotics, these intermediate points are often called via points, since whatever path the end-effector actually takes, it will definitely go via these intermediate points. The placement of these points can be an important consideration when planning a path which is intended to avoid obstacles in the robot's work space.

To illustrate this approach we will look again at the straight line example above, but this time using two segments. The segments will run from $t = 0$ to $t = 1/2$ and from $t = 1/2$ to $t = 1$. When $t = 1/2$ the co-ordinates of the wrist centre should be:-

$$\mathbf{p}_c(1/2) = 1/2 \begin{pmatrix} 5/\sqrt{2} \\ -3/\sqrt{2} \\ -4 \end{pmatrix} + 1/2 \begin{pmatrix} -(2 + \frac{3}{\sqrt{2}}) \\ -(2 + \frac{5}{\sqrt{2}}) \\ \frac{4}{\sqrt{2}} \end{pmatrix}$$

The inverse kinematics will give us the values for the joint angles here:-

$$\boldsymbol{\theta}(1/2) = \begin{pmatrix} -0.3398 \\ 0.6450 \\ 2.1650 \end{pmatrix}$$

In the first segment we set:-

$$\boldsymbol{\theta}_l(t) = 2(\frac{1}{2} - t)\boldsymbol{\theta}(0) + 2t\boldsymbol{\theta}(\frac{1}{2})$$

and in the second segment we put:-

$$\boldsymbol{\theta}_l(t) = 2(1 - t)\boldsymbol{\theta}(\frac{1}{2}) + t\boldsymbol{\theta}(1)$$

See fig. 7.2. Notice that the error is indeed smaller, the trajectory of the wrist centre is closer to the target line. However, at the midpoint there is a discontinuity in the velocity, this is very undesirable.

In general, if we have an interval $[a, b]$ and values $f(a)$ and $f(b)$ at the end points the linearly interpolation function on the interval is given by:-

$$f_{lin}(t) = \frac{(t - b)}{a - b}f(a) - \frac{(t - a)}{a - b}f(b), \qquad a \le t \le b$$

The idea is that the function is linear in t, and when $t = a$ the coefficient of $f(b)$ vanishes and the function reduces to $f(a)$. Then when $t = b$, the coefficient of $f(a)$ disappears and the function becomes $f(b)$. So the function is linear on the interval and has the designated values at the end points.

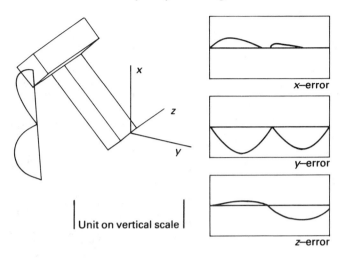

Figure 7.2 Approximate Straight Line using Two Linearly Interpolated Segments

7.3   Polynomial Approximations

For a more sophisticated approximation we use higher degree polynomials. Polynomials can be evaluated very rapidly: computing a degree n polynomial requires only n multiplications, using Horner's method. For this we rewrite a degree n polynomial using nested multiplication:-

$$a_n t^n + a_{n-1} t^{n-1} + \cdots + a_1 t^1 + a_0 = (\cdots((a_n t + a_{n-1})t + a_{n-2})t \cdots a_1)t + a_0$$

As an example let us compute the quadratic approximation to the straight line path of the previous section. The general quadratic is of the form:-

$$\boldsymbol{\theta}_q(t) = \mathbf{a}t^2 + \mathbf{b}t + \mathbf{c}$$

The coefficients have been written as vectors, since actually we have a polynomial for each joint. Anything we do to find the coefficients will have to be done for each joint, but as the procedures will be the same for each joint the vector notation can be used to summarize the results.

The quadratic must pass through the three points $\boldsymbol{\theta}(0), \boldsymbol{\theta}(1/2)$ and $\boldsymbol{\theta}(1)$. This gives us three equations which we can use to find the three constants:-

$$\begin{aligned}
\boldsymbol{\theta}(0) &= \mathbf{c} \\
\boldsymbol{\theta}(1/2) &= \mathbf{a}/4 + \mathbf{b}/2 + \mathbf{c} \\
\boldsymbol{\theta}(1) &= \mathbf{a} + \mathbf{b} + \mathbf{c}
\end{aligned}$$

This has the solution:-

$$\begin{aligned}
\mathbf{a} &= 2\boldsymbol{\theta}(1) - 4\boldsymbol{\theta}(1/2) + 2\boldsymbol{\theta}(0) \\
\mathbf{b} &= -\boldsymbol{\theta}(1) + 4\boldsymbol{\theta}(1/2) - 3\boldsymbol{\theta}(0)
\end{aligned}$$

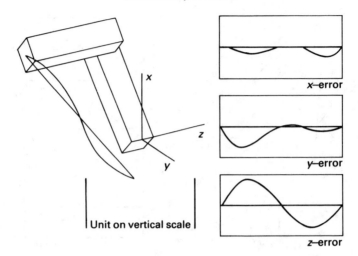

Figure 7.3 Quadratic Approximation to Straight Line Path

$$\mathbf{c} = \boldsymbol{\theta}(0)$$

Really, we have nine equations here since the coefficients are vectors. In this case the quadratic approximation gives:-

$$\boldsymbol{\theta}_q = \begin{pmatrix} 1.3592 \\ 2.1324 \\ -3.9476 \end{pmatrix} t^2 + \begin{pmatrix} -2.9300 \\ -2.9178 \\ 3.1622 \end{pmatrix} t + \begin{pmatrix} -0.7854 \\ 0.7854 \\ 0.7854 \end{pmatrix}$$

See fig. 7.3.

In general we get better accuracy with a higher degree polynomial. For a degree n approximation:-

$$\boldsymbol{\theta}_n(t) = \mathbf{c}_n t^n + \mathbf{c}_{n-1} t^{n-1} + \cdots + \mathbf{c}_2 t^2 + \mathbf{c}_1 t + \mathbf{c}_0$$

we need to find the $n+1$ constant vectors $\mathbf{c}_i$. This means we need $n+1$ via points $\boldsymbol{\theta}(t_i)$. These points do not have to be regularly spaced, but they do have to be distinct. Given these $n+1$ via points, to find the degree n polynomial which passes through these points we must solve the following system of linear equations:-

$$\begin{pmatrix} t_0^n & t_0^{n-1} & \cdots & t_0 & 1 \\ t_1^n & t_1^{n-1} & \cdots & t_1 & 1 \\ \vdots & \vdots & \ddots & \vdots & \vdots \\ t_n^n & t_n^{n-1} & \cdots & t_n & 1 \end{pmatrix} \begin{pmatrix} \mathbf{c}_n \\ \mathbf{c}_{n-1} \\ \vdots \\ \mathbf{c}_0 \end{pmatrix} = \begin{pmatrix} \boldsymbol{\theta}(t_n) \\ \boldsymbol{\theta}(t_{n-1}) \\ \vdots \\ \boldsymbol{\theta}(t_0) \end{pmatrix}$$

This is a classical problem in numerical analysis, usually known as Lagrange interpolation, and it has an analytic solution. The Lagrange interpolated polynomial would have the form:-

$$\boldsymbol{\theta}_L(t) = \sum_{i=0}^{n} \boldsymbol{\theta}_i \frac{(t - t_0)(t - t_1) \cdots (t - t_{i-1})(t - t_{i+1}) \cdots (t - t_n)}{(t_i - t_0)(t_i - t_1) \cdots (t_i - t_{i-1})(t_i - t_{i+1}) \cdots (t_i - t_n)}$$

Notice that each term in the sum is a degree n polynomial in t. For some via point t_i all the terms in the sum disappear except one, that which does not contain the term $(t - t_i)$. This term simplifies to just $\boldsymbol{\theta}_i$. Hence for each via point we have:-

$$\boldsymbol{\theta}_L(t_i) = \boldsymbol{\theta}_i$$

as required.

High degree polynomial approximations become increasingly wiggly. Although the error may be small, this is undesirable when trying to follow a smooth path, since resonant vibrations of the robot's links and joints can be excited. Also the higher the degree, the more computation is needed. In general it is uncommon to use anything higher than cubic polynomials, but sometimes up to degree five polynomials are used.

We should really say something about the error produced by these approximations. A great deal is known about the errors made in Lagrange interpolation. The problem was studied at the end of the last century, by the Russian mathematician Chebyshev. His interest in this problem stemmed from his study of mechanisms, especially the problem of drawing curves mechanically; this is usually forgotten by numerical analysis texts. Chebyshev found where to place the interpolation point so as to minimize the error. Unfortunately this result is not very useful for us here, since the error minimized would be the joint space error. That is the difference between the interpolation polynomial and the desired function. The important error for us would be the error in work space, that is the error in the position and orientation of the robot's end-effector. A reasonable rule of thumb would be to space the via points more closely when the determinant of the jacobian is large, since in these positions small joint errors will produce large errors in position.

However, in real robots the sensors on the joints only have a finite accuracy. Typically an angular encoder will have 1,000 divisions per revolution. After a gear reduction of 10 to 1 we will not be able to distinguish joint angles differing by about 10^{-4} radians. So there is no point in getting the joint space error any smaller than this.

The error made by using a degree n Lagrange interpolated polynomial, $\theta_L(t)$, to approximate a function $\theta(t)$ on the interval $[0, 1]$, is given by:-

$$|\theta_L(t) - \theta(t)| \le \frac{M_{n+1}}{(n + 1)!} |(t - t_0)(t - t_1) \cdots (t - t_n)|$$

The quantity M_{n+1} depends only on the original function:-

$$M_{n+1} = \max_{0 \le \xi \le 1} \left| \frac{d^{n+1}\theta(\xi)}{dt^{n+1}} \right|$$

If this quantity can be estimated for the joint angles, the maximum error can be predicted.

In the above formula for the error, the term $|(t - t_0)(t - t_1) \cdots (t - t_n)|$ only depends on the position of the interpolation points. Hence a good choice of via points will lead to

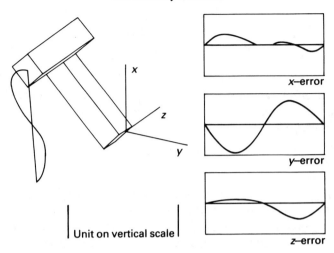

Figure 7.4 Linear and Quadratic Matched Segments

a more accurate approximation. The best possible choice for these points is at the zeros of the degree $n + 1$ Chebyshev polynomial. They are given by:-

$$t_i = \frac{1}{2} + \frac{1}{2} \cos \frac{(2i + 1)\pi}{(2n + 2)} \qquad\qquad i = 0, 1, \ldots n$$

Notice that these points are not evenly spaced, they bunch up towards the ends of the interval.

7.4   Matching Derivatives

One way of avoiding the use of high degree polynomials is to use several segments, just like the several linear approximations we saw in the last section. However, we do not want the discontinuity in velocity we saw there. This can be done by matching the derivatives at the end points of the segment. To illustrate this we will look at the straight line example above again. Suppose in the first segment we use a linear approximation exactly as in section 7.2:-

$$\boldsymbol{\theta}_l(t) = 2(\frac{1}{2} - t)\boldsymbol{\theta}(0) + 2t\boldsymbol{\theta}(\frac{1}{2})$$

Now in the second segment we have the following three constraints on the polynomial; the polynomial must pass through the two points:-

$$\boldsymbol{\theta}(1/2) \qquad \text{and} \qquad \boldsymbol{\theta}(1)$$

The third constraint is that the derivative of the polynomial at $t = 1/2$ is the same as the derivative of the preceding approximation at this point. This is given by:-

$$\dot{\theta}_l(1/2) = -2\theta(0) + 2\theta(1/2) = 2\theta(\frac{1}{2}) - 2\theta(0)$$

Since we have three pieces of information we will need a quadratic polynomial because such polynomials have three parameters:-

$$\theta_q(t) = \mathbf{a}t^2 + \mathbf{b}t + \mathbf{c}$$

The three bits of information give us three equations for the coefficients:-

$$\begin{array}{rclcl}
\theta_q(1/2) & = & \frac{1}{4}\mathbf{a} + \frac{1}{2}\mathbf{b} + \mathbf{c} & = & \theta(1/2) \\
\theta_q(1) & = & \mathbf{a} + \mathbf{b} + \mathbf{c} & = & \theta(1) \\
\dot{\theta}_q(1/2) & = & \mathbf{a} + \mathbf{b} & = & 2\theta(\frac{1}{2}) - 2\theta(0)
\end{array}$$

These equations are easily solved to give:-

$$\mathbf{a} = 4\theta(1) - 8\theta(1/2) + 4\theta(0) = \begin{pmatrix} 2.718 \\ 4.265 \\ -7.895 \end{pmatrix}$$

$$\mathbf{b} = -4\theta(1) + 10\theta(1/2) - 6\theta(0) = \begin{pmatrix} 4.456 \\ 12.733 \\ 27.933 \end{pmatrix}$$

$$\mathbf{c} = \theta(1) - 2\theta(1/2) + 2\theta(0) = \begin{pmatrix} 1.465 \\ 2.637 \\ -0.403 \end{pmatrix}$$

The path they describe is illustrated in fig. 7.4.

To specify a polynomial we may require it to pass through certain points or to have a given slope at specific points. Combinations of both are possible and it is also possible to look at higher derivatives too. This more general type of interpolation is known as Hermite interpolation in the literature. The example given above uses only linear and quadratic polynomials; however, it is more usual to use cubic polynomials as segments. These are sometimes called splines. The name comes from a mechanical curve fitting device which uses a stiff wire to represent the curve.

Now we turn to the lift-off and set-down phases of trajectories. Using polynomial approximations we can say something about the velocities. Recall that for the lift-off phase of a trajectory the joints must begin with zero speed. This can be achieved with a quadratic polynomial θ_b:-

$$\theta_b(t) = \mathbf{a}t^2 + \mathbf{b}t + \mathbf{c}$$

The joint velocities are now given by the derivative:-

$$\dot{\theta}_b(t) = 2\mathbf{a}t + \mathbf{b}$$

and so at the beginning of the time interval we have:-

$$\dot{\theta}_b(0) = \mathbf{b} = 0$$

So the coefficients **b** must be zero, similarly, the coefficient **c** is the starting position, $c = \theta(0)$. We could find the **a** coefficient by requiring $\theta_b(1)$ to be on the desired trajectory. Alternatively we could demand that the velocity at the end point of the time interval should be along the desired path. However, it is more usual to find **a** by choosing the acceleration of the path; for a quadratic polynomial the joint accelerations are constant:-

$$\ddot{\theta}_b(t) = 2\mathbf{a}$$

In a real robot the motors will have maximum accelerations and also maximum speeds. We can run the robot at maximum acceleration until we get the maximum velocity. Usually the robot will accelerate and decelerate along straight line segments. So we know the direction along which the acceleration must occur. We can increase the magnitude of the acceleration until the acceleration of one of the joints becomes equal to its maximum value. We will call the resulting vector of joint accelerations **A**, and the maximum of the joint accelerations, that is the largest component of **A**, will be labelled A_{max}. The joint which is accelerating most quickly will reach its limiting angular velocity, ω_{max}, after a time t_1; where:-

$$\omega_{max} = 2A_{max}t_1$$

So this acceleration phase must end at time $t_1 = \omega_{max}/A_{max}$. The quadratic is thus given by:-

$$\theta_b(t) = \frac{1}{2}\mathbf{A}t^2 + \theta(0) \qquad\qquad 0 \le t \le t_1$$

Quadratic polynomials can also be used for the set-down stage of the trajectory. Now we want the velocities to vanish at the end of the path:-

$$\theta_e(1) = \theta(1) \qquad \text{and} \qquad \dot{\theta}_e(1) = 0$$

The final piece of information we need is the deceleration for the path, which we treat exactly as we did for the acceleration. Let $\ddot{\theta}_e(t) = \mathbf{D}$ be the acceleration vector for the path and D_{max} the acceleration of the most rapidly decelerating joint. Not only might this be a different joint from the one that was accelerating most quickly in the lift-off phase, but also the maximum deceleration the motors can supply is not necessarily just the negative of the maximum acceleration.

With this information we can find the three coefficients of the quadratic polynomial:-

$$\theta_e(t) = \frac{1}{2}\mathbf{D}t^2 - \mathbf{D}t + \frac{1}{2}\mathbf{D} + \theta(1) \qquad\qquad t_2 \le t \le 1$$

The beginning of the set-down phase, t_2, can be found in the same way that we found the end of the lift-off phase. At t_2 the quadratic will give the maximum velocity of the most rapidly decelerating joint ω'_{max}. This time is given by $t_2 = (\omega'_{max} + D_{max})/D_{max}$.

For the middle section of the trajectory we could use any of the techniques mentioned in the previous sections. Often, one or more cubic polynomials are used. We would need to use the inverse kinematics to find intermediate points along the path. So here we will just look at using a single cubic:-

$$\theta(t) = \mathbf{a}t^3 + \mathbf{b}t^2 + \mathbf{c}t + \mathbf{d}$$

To find the four coefficients we need four pieces of information, and these are given by the positions and velocities at the two times t_1 and t_2:-

$$
\begin{array}{rcccl}
\boldsymbol{\theta}_c(t_1) & = & \boldsymbol{\theta}_b(t_1) & = & \frac{1}{2}\mathbf{A}t_1^2 + \boldsymbol{\theta}(0) \\
\boldsymbol{\theta}_c(t_2) & = & \boldsymbol{\theta}_{q2}(t_2) & = & \frac{1}{2}\mathbf{D}t_2^2 - \mathbf{D}t_2 + \frac{1}{2}\mathbf{D} + \boldsymbol{\theta}(1) \\
\dot{\boldsymbol{\theta}}_c(t_1) & = & \dot{\boldsymbol{\theta}}_b(t_1) & = & \mathbf{A}t_1 \\
\dot{\boldsymbol{\theta}}_c(t_2) & = & \dot{\boldsymbol{\theta}}_{q2}(t_2) & = & \mathbf{D}t_2
\end{array}
$$

Solving this system of linear equations gives cubic polynomials which connect with the end points of the lift-off and set-down phases, and, moreover, the derivatives will be continuous at the end points. The approximate trajectory will be given by:-

$$
\boldsymbol{\theta}_{approx}(t) = \left\{
\begin{array}{lll}
\boldsymbol{\theta}_b(t), & \text{if} & 0 \le t \le t_1 \\
\boldsymbol{\theta}_c(t), & \text{if} & t_1 \le t \le t_2 \\
\boldsymbol{\theta}_e(t), & \text{if} & t_2 \le t \le 1
\end{array}
\right.
$$

To conclude, there are many different ways to approximate the joint trajectories for any desired end-effector trajectory. The choice of any particular method will depend on such factors as the accuracy needed and the time available for the computations.

Exercises

7.1 A two joint planar manipulator has link lengths $l_1 = 2$ units and $l_2 = 1$ unit. The end point of the manipulator is required to travel with constant velocity from the point $(1/2+\sqrt{3}, -1-\sqrt{3}/2)$ to the point $(1/2+\sqrt{3}, 1+\sqrt{3}/2)$, along the straight line $x = 1/2 + \sqrt{3}$. Find:-

    (i) The linear approximation to the path in joint space which produces this trajectory.

    (ii) The quadratic approximation for the trajectory which also passes through the point $(1/2 + \sqrt{3}, 0)$.

Use the elbow down posture.

7.2 Consider the three joint wrist introduced in section 4.2. It is required that the end-effector rotates uniformly about the x-axis, starting from the position where $(\theta_1, \theta_2, \theta_3) = (0, \pi/4, 0)$. The path can be described as:-

$$
\mathbf{K}(t) = \begin{pmatrix} 1 & 0 & 0 \\ 0 & \cos \pi t & -\sin \pi t \\ 0 & \sin \pi t & \cos \pi t \end{pmatrix} \begin{pmatrix} \frac{1}{\sqrt{2}} & 0 & \frac{1}{\sqrt{2}} \\ 0 & 1 & 0 \\ -\frac{1}{\sqrt{2}} & 0 & \frac{1}{\sqrt{2}} \end{pmatrix} \qquad 0 \le t \le 1
$$

Find the linear approximation to the path in joint space and the quadratic approximation which passes through a via point at $t = 1/2$.

7.3   Find a degree four polynomial approximation to the uniform straight line trajectory described in section 7.2. The path must pass through the three points:-

$$\boldsymbol{\theta}(0) = \begin{pmatrix} \pi/4 \\ \pi/2 \\ \pi/2 \end{pmatrix}, \qquad \boldsymbol{\theta}(1/2) = \begin{pmatrix} -0.3398 \\ 0.6450 \\ 2.1650 \end{pmatrix}, \qquad \boldsymbol{\theta}(1) = \begin{pmatrix} -\pi/4 \\ \pi/4 \\ \pi/4 \end{pmatrix}$$

Also the path must have the same velocities in work space as the straight line path at the points $\boldsymbol{\theta}(0)$ and $\boldsymbol{\theta}(1)$.

7.4   It well known in numerical analysis that it is often more economical to interpolate functions using rational approximations; that is, a function that is a ratio of two polynomials. Find the rational approximation of the form:-

$$\theta_i(t) \approx \frac{a_i t + b_i}{c_i t + 1} \qquad\qquad i = 1, 2, 3$$

which approximates the same path as in exercise 7.3, but is only required to pass through the same three points.

8 Statics

8.1 Forces and Torques

In this chapter we introduce forces, but only static ones. Later we will see how to deal with moving forces and dynamics. For the moment we content ourselves with static forces and rigid bodies in equilibrium.

Forces are vector quantities. For a rigid body to be in equilibrium the vector sum of the forces acting on it must be zero. However, this does not ensure equilibrium: the sum of the moments of the forces must equal zero. Forces act along lines, and crudely speaking the moment of a force about a point O is the perpendicular distance from O to the line multiplied by the magnitude of the force. In fact the moment of a force is also a vector quantity. Suppose $\mathbf{F}$ is a force vector, and $\mathbf{r}$ is the position vector of any point on the line of action of the force. Then the moment of the force about the origin is given by the vector product:-

$$\mathbf{M} = \mathbf{r} \wedge \mathbf{F}$$

Notice that we may take $\mathbf{r}$ as any point on the line, since any other point will be of the form $\mathbf{r}' = \mathbf{r} + \lambda \mathbf{F}$. So this would give a moment:-

$$\mathbf{r}' \wedge \mathbf{F} = (\mathbf{r} + \lambda \mathbf{F}) \wedge \mathbf{F} = \mathbf{r} \wedge \mathbf{F}$$

because $\mathbf{F} \wedge \mathbf{F} = \mathbf{0}$.

We may combine the force and moment into a single six dimensional vector $\begin{pmatrix} \mathbf{M} \\ \mathbf{F} \end{pmatrix}$. Do not worry too much that the units for the last three components have dimensions of force while the first three have dimensions force × length; we will not do anything which will mix quantities with different dimensions. Such a 'force-moment' vector is called a **wrench**. Now we may say that a rigid body is in equilibrium if and only if the total wrench acting on it is zero.

Wrenches are very similar to the instantaneous screws we met in section 6.6. Like instantaneous screws wrenches have a pitch associated with them. For a wrench the ratio $\mathbf{M} \cdot \mathbf{F} : \mathbf{F} \cdot \mathbf{F}$ is called the pitch of the wrench. If the pitch is $0 : 1$, then the wrench is a

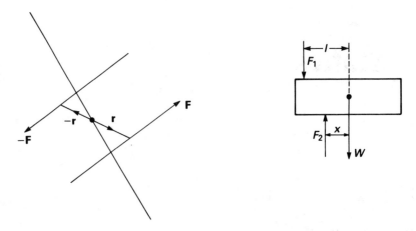

Figure 8.1 (a) A Couple of Forces (b) Forces for Holding a Rectangular Block

pure force. This is easy to see, since if the wrench is a pure force the moment has the form $\mathbf{M} = \mathbf{r} \wedge \mathbf{F}$ and hence $\mathbf{M} \cdot \mathbf{F} = (\mathbf{r} \wedge \mathbf{F}) \cdot \mathbf{F} = 0$, using the cyclic properties of the scalar triple product.

It is possible to have wrenches with other pitches. For example, consider a couple of forces acting on a rigid body as illustrated in fig. 8.1(a). The total wrench acting on the body is given by the vector sum:-

$$\begin{pmatrix} \mathbf{r} \wedge \mathbf{F} \\ \mathbf{F} \end{pmatrix} + \begin{pmatrix} (-\mathbf{r}) \wedge (-\mathbf{F}) \\ -\mathbf{F} \end{pmatrix} = \begin{pmatrix} 2\mathbf{r} \wedge \mathbf{F} \\ 0 \end{pmatrix}$$

Such a wrench represents a pure torque and its pitch is (by definition) $1 : 0$.

In between we may have wrenches of pitch $\alpha : \beta$. In general a wrench is a combination of a force and a torque about the same axis as the force. So as with the instantaneous screws, wrenches are associated to lines in space; the axis of the wrench. A general wrench has the form:-

$$W = \begin{pmatrix} \mathbf{r} \wedge \mathbf{F} + p\mathbf{F} \\ \mathbf{F} \end{pmatrix}$$

where $p = \alpha/\beta$ is the pitch of the wrench, and $\mathbf{r}$ is the position vector of any point on the axis.

8.2 Gripping

An immediate application of these ideas is to gripping problems. At the moment most robot grippers are rather basic in design. Usually they consist of two fingers which are only able to move parallel to each other. For many industrial applications such designs are quite adequate. However, dextrous multi-fingered grippers are under development.

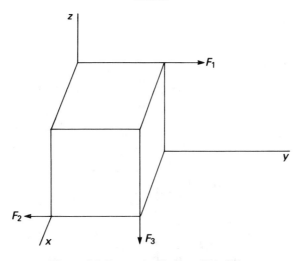

Figure 8.2 Forces Applied to a Unit Cube

Imagine trying to hold a rectangular block with two fingers, see fig. 8.1(b). The weight W of the block acts through the centre of mass and the finger forces F_1 and F_2 are separated by a small distance x. To balance the forces we must satisfy:-

$$-F_1 + F_2 = W$$

and to balance the moments we need:-

$$(l + x)F_1 - lF_2 = 0$$

Here we have taken moments about the centre of mass. These are two simultaneous linear equations in the unknown finger forces. Their solution is easily found:-

$$F_1 = \frac{(l - x)}{x}W, \qquad F_2 + \frac{l}{x}W$$

Notice that if the distance x was zero then we could only balance the moments by gripping the block opposite the centre of mass. This is one reason why grippers normally make contact with the workpiece via flat 'cheeks'. Another reason, though, is to increase the friction between the gripper and the workpiece.

If the forces do not all lie in the same plane, then we need to use the concept of wrenches introduced above. To see how this works we will look at an example. Suppose a robot gripper applies forces to a unit cube as shown in fig. 8.2. Let us find the total wrench acting on the cube when the magnitudes of the forces are:-

$$F_1 = 1, \quad F_2 = 2, \quad F_3 = 1$$

in some units.

Let the wrenches be W_1, W_2 and W_3, then:-

$$W_1 = \begin{pmatrix} -F_1 \\ 0 \\ 0 \\ 0 \\ F_1 \\ 0 \end{pmatrix}, \qquad W_2 = \begin{pmatrix} 0 \\ 0 \\ -F_2 \\ 0 \\ -F_2 \\ 0 \end{pmatrix}, \qquad W_3 = \begin{pmatrix} -F_3 \\ F_3 \\ 0 \\ 0 \\ 0 \\ -F_3 \end{pmatrix}$$

since:-

$$\begin{pmatrix} 0 \\ 0 \\ 1 \end{pmatrix} \wedge \begin{pmatrix} 0 \\ F_1 \\ 0 \end{pmatrix} = \begin{pmatrix} -F_1 \\ 0 \\ 0 \end{pmatrix},$$

$$\begin{pmatrix} 1 \\ 0 \\ 0 \end{pmatrix} \wedge \begin{pmatrix} 0 \\ -F_2 \\ 0 \end{pmatrix} = \begin{pmatrix} 0 \\ 0 \\ -F_2 \end{pmatrix},$$

$$\begin{pmatrix} 1 \\ 1 \\ 0 \end{pmatrix} \wedge \begin{pmatrix} 0 \\ 0 \\ -F_3 \end{pmatrix} = \begin{pmatrix} -F_3 \\ F_3 \\ 0 \end{pmatrix}$$

So the total wrench is:-

$$W_1 + W_2 + W_3 = \begin{pmatrix} -F_1 \\ 0 \\ 0 \\ 0 \\ F_1 \\ 0 \end{pmatrix} + \begin{pmatrix} 0 \\ 0 \\ -F_2 \\ 0 \\ -F_2 \\ 0 \end{pmatrix} + \begin{pmatrix} -F_3 \\ F_3 \\ 0 \\ 0 \\ 0 \\ -F_3 \end{pmatrix} = \begin{pmatrix} -(F_1 + F_3) \\ F_3 \\ -F_2 \\ 0 \\ (F_1 - F_2) \\ -F_3 \end{pmatrix} = \begin{pmatrix} -2 \\ 1 \\ -2 \\ 0 \\ -1 \\ -1 \end{pmatrix}$$

What are the pitch and axis of this wrench?

The pitch of this wrench is given by:-

$$((F_1 - F_2)F_3 + F_2 F_3)/((F_1 - F_2)^2 + F_3^2) = F_1 F_3/((F_1 - F_2)^2 + F_3^2) = \frac{1}{2}$$

To find the axis of the wrench we need the direction of the axis and a point on the axis. The unit vector in the direction of the axis is just:-

$$\hat{\mathbf{F}} = \frac{1}{\sqrt{2}} \begin{pmatrix} 0 \\ -1 \\ -1 \end{pmatrix}$$

A point on the axis can be found from the first three components of the wrench.

$$\begin{pmatrix} -2 \\ 1 \\ -2 \end{pmatrix} = \mathbf{r} \wedge \begin{pmatrix} 0 \\ -1 \\ -1 \end{pmatrix} + p \begin{pmatrix} 0 \\ -1 \\ -1 \end{pmatrix}$$

Rearranging gives:-

$$\mathbf{r} \wedge \begin{pmatrix} 0 \\ -1 \\ -1 \end{pmatrix} = \begin{pmatrix} -2 \\ 3/2 \\ -3/2 \end{pmatrix}$$

This gives the three equations:-

$$-r_y + r_z = -2, \qquad r_x = 3/2, \qquad -r_x = -3/2$$

Certainly this gives us that $r_x = 3/2$ but we cannot solve for r_y and r_z. Remember for any solution $\mathbf{r}$, to these equations there are always other solutions $\mathbf{r} + \lambda\mathbf{F}$. To get a unique solution we can impose the extra condition $\mathbf{r} \cdot \mathbf{F} = 0$, that is the position vector of the point on the axis must be perpendicular to the axis. The result is two equations:-

$$-r_y + r_z = -2, \qquad -r_y - r_z = 0$$

These can now be solved to give:-

$$\mathbf{r} = \begin{pmatrix} 3/2 \\ 1 \\ -1 \end{pmatrix}$$

This leads to an interesting question. How many fingers does a gripper need in general? Consider gripping a solid object without friction. Each finger applies a force normal to the surface of the object. To hold the object we must balance the object's weight and any other disturbing wrenches. If we assume the finger positions are fixed then we have to be able to solve for the magnitude of the finger forces. This means that six fingers are needed, to ensure that there are the same number of variables as degrees-of-freedom for the object. For a gripper with six fingers we would have to solve the following system of linear equations:-

$$\begin{pmatrix} \mathbf{r}_1 \wedge \mathbf{f}_1 & \mathbf{r}_2 \wedge \mathbf{f}_2 & \mathbf{r}_3 \wedge \mathbf{f}_3 & \mathbf{r}_4 \wedge \mathbf{f}_4 & \mathbf{r}_5 \wedge \mathbf{f}_5 & \mathbf{r}_6 \wedge \mathbf{f}_6 \\ \mathbf{f}_1 & \mathbf{f}_2 & \mathbf{f}_3 & \mathbf{f}_4 & \mathbf{f}_5 & \mathbf{f}_6 \end{pmatrix} \begin{pmatrix} F_1 \\ F_2 \\ F_3 \\ F_4 \\ F_5 \\ F_6 \end{pmatrix} = \begin{pmatrix} \mathbf{M} \\ \mathbf{W} \end{pmatrix}$$

Here $\mathbf{W}$ and $\mathbf{M}$ are the weight and moment of the weight. The F_i are the magnitudes of the finger forces, so we must normalize the columns of the coefficient matrix so that $\mathbf{f}_i \cdot \mathbf{f}_i = 1$ for every finger. Whatever the right-hand side is, we can always solve this equation for the magnitudes of the finger forces, unless the determinant of the coefficient matrix is zero. However, the coefficient matrix depends only on the finger positions, and hence the matrix being singular corresponds to a 'bad' choice of finger positions.

On some surfaces every choice of finger positions is 'bad'. Such surfaces are in fact exactly the surfaces of the Reuleaux lower pairs that we met in section 3.2. These objects cannot be grasped effectively; in other words we cannot completely constrain the motion of the lower Reuleaux pairs without friction. The real use of this result is that it tells us

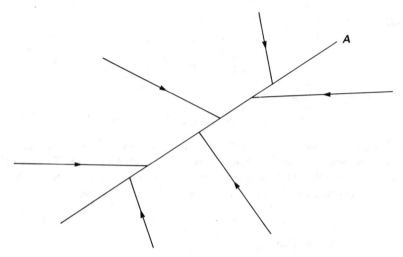

Figure 8.3 A 'Bad' Choice of Finger Positions

what 'bad' choices of finger position look like. For example, suppose the six lines of force all met some other line, A, in fig. 8.3. Now there will be some surface of revolution with axis A such that all the force lines are surface normals. So this is a 'bad' choice of finger positions; without friction it is impossible to balance a torque about the axis A, and so we will not be able to stop the object rotating about A.

Similarly if all the lines are perpendicular to some direction, there will be a surface of translation normal to all of them. Hence, forces along the given direction will not be able to be cancelled.

In reality there is friction, and if we are prepared to use friction to help us grip objects then only two or three fingers are necessary. This works very well for human hands for example, but be careful when trying to hold a bar of soap!

8.3 Duality Between Wrenches and Screws

In chapter 6 we met another kind of six dimensional vector, the instantaneous screw. Here we show that there is a close relationship between wrenches and instantaneous screws: the relationship is in the form of a 'duality'. We begin by considering an arbitrary rigid body acted on by a wrench $\mathcal{W}$. This wrench causes the body to acquire an instantaneous screw motion S. The instantaneous power exerted by the wrench is

$$\mathcal{W}^T \mathbf{S} = (\mathbf{M}^T, \mathbf{F}^T) \begin{pmatrix} \omega \\ \mathbf{u} \wedge \omega + p\omega \end{pmatrix} = \mathbf{M} \cdot \omega + \mathbf{F} \cdot (\mathbf{u} \wedge \omega) + p\mathbf{F} \cdot \omega$$

This generalizes the expressions for power from ordinary mechanics; $\mathbf{F} \cdot \mathbf{v}$ for linear motion, and $\mathbf{M} \cdot \omega$ for rotational motion. Thus we have a pairing operation between wrenches and instantaneous screws, the result of which is a scalar; the power.

Given a set of six linearly independent wrenches $\{\mathcal{W}_1, \mathcal{W}_2, \mathcal{W}_3, \mathcal{W}_4, \mathcal{W}_5, \mathcal{W}_6\}$, we can always find a **dual** set of six linearly independent instantaneous screws $\{\mathbf{S}_1, \mathbf{S}_2, \mathbf{S}_3, \mathbf{S}_4, \mathbf{S}_5, \mathbf{S}_6\}$, such that:-

$$\mathcal{W}_i^T \mathbf{S}_j = \begin{cases} 0 & \text{if } i \neq j \\ 1 & \text{if } i = j \end{cases}$$

For example, suppose the six wrenches consist of three unit forces along three orthogonal lines:-

$$\mathcal{W}_1 = \begin{pmatrix} \mathbf{r} \wedge \mathbf{i} \\ \mathbf{i} \end{pmatrix}, \qquad \mathcal{W}_2 = \begin{pmatrix} \mathbf{r} \wedge \mathbf{j} \\ \mathbf{j} \end{pmatrix}, \qquad \mathcal{W}_3 = \begin{pmatrix} \mathbf{r} \wedge \mathbf{k} \\ \mathbf{k} \end{pmatrix}$$

together with the unit torques about these directions:-

$$\mathcal{W}_4 = \begin{pmatrix} \mathbf{i} \\ \mathbf{0} \end{pmatrix}, \qquad \mathcal{W}_5 = \begin{pmatrix} \mathbf{j} \\ \mathbf{0} \end{pmatrix}, \qquad \mathcal{W}_6 = \begin{pmatrix} \mathbf{k} \\ \mathbf{0} \end{pmatrix}$$

To find the dual set of screws we form the matrix where the columns are the wrenches:-

$$\mathbf{M} = \begin{pmatrix} \mathcal{W}_1 & \mathcal{W}_2 & \mathcal{W}_3 & \mathcal{W}_4 & \mathcal{W}_5 & \mathcal{W}_6 \end{pmatrix}$$

The rows of the inverse of this matrix are then the transposes of the dual screws:-

$$\mathbf{M}^{-1} = \begin{pmatrix} \mathbf{S}_1^T \\ \mathbf{S}_2^T \\ \mathbf{S}_3^T \\ \mathbf{S}_4^T \\ \mathbf{S}_5^T \\ \mathbf{S}_6^T \end{pmatrix}$$

This works because the relation $\mathbf{M}^{-1}\mathbf{M} = \mathbf{I}$ summarizes the duality conditions given above. The fact that the wrenches must be linearly independent guarantees that the inverse matrix exists, since $\det \mathbf{M} \neq 0$ if and only if the wrenches are linearly independent.

For our example we do not have to perform the matrix inversion since it is simple to check that the dual screws are given by:-

$$\mathbf{s}_1 = \begin{pmatrix} \mathbf{0} \\ \mathbf{i} \end{pmatrix}, \qquad \mathbf{s}_2 = \begin{pmatrix} \mathbf{0} \\ \mathbf{j} \end{pmatrix}, \qquad \mathbf{s}_3 = \begin{pmatrix} \mathbf{0} \\ \mathbf{k} \end{pmatrix}$$

and:-

$$\mathbf{s}_4 = \begin{pmatrix} \mathbf{i} \\ \mathbf{r} \wedge \mathbf{i} \end{pmatrix}, \qquad \mathbf{s}_5 = \begin{pmatrix} \mathbf{j} \\ \mathbf{r} \wedge \mathbf{j} \end{pmatrix}, \qquad \mathbf{s}_6 = \begin{pmatrix} \mathbf{k} \\ \mathbf{r} \wedge \mathbf{k} \end{pmatrix}$$

Now an arbitrary wrench $\mathcal{W}$ can always be written in terms of a set of six linearly independent wrenches:-

$$\mathcal{W} = a_1 \mathcal{W}_1 + a_2 \mathcal{W}_2 + a_3 \mathcal{W}_3 + a_4 \mathcal{W}_4 + a_5 \mathcal{W}_5 + a_6 \mathcal{W}_6$$

The a_i are constant coefficients. These coefficients can be found by simply pairing with the ith dual screw.

$$a_i = \mathcal{W}^T \mathbf{s}_i$$

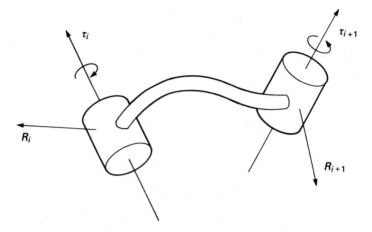

Figure 8.4 The Forces and Torques acting on a General Link

This is very similar to using the dot product to project out components of three dimensional vectors.

This has immediate applications in robotics. Consider a robot with six revolute joints arranged serially. If torques are applied to the joints what is the resulting wrench on the end-effector? We just want to know the effects of the joint torques so we will ignore the weight of the links and dynamic effects which will also give rise to wrenches at the end-effector.

Consider the forces and torques acting on a single link, see fig. 8.4. At each joint we have a torque from the actuator and a reaction wrench caused by the adjacent link. For equilibrium we must have:-

$$\boldsymbol{\tau}_i + \mathcal{R}_i - \boldsymbol{\tau}_{i+1} - \mathcal{R}_{i+1} = 0$$

except at the final link where we have:-

$$\boldsymbol{\tau}_6 + \mathcal{R}_6 - \mathcal{W} = 0$$

Here $\mathcal{W}$ is the wrench at the end-effector that we are trying to find. Manipulating the above equations we get six equations of the form:-

$$\boldsymbol{\tau}_i + \mathcal{R}_i = \mathcal{W}$$

Let us separate out the magnitude of the actuator torques by writing them as $\boldsymbol{\tau}_i = \tau_i \begin{pmatrix} \hat{\mathbf{v}}_i \\ 0 \end{pmatrix}$. Now consider the wrenches on the ith joint. We choose six linearly independent wrenches, as in the example above, but one must be a unit torque about the joint. So now we can find the magnitude of the joint torque by pairing with the dual screw, $\mathbf{S}_i = \begin{pmatrix} \hat{\mathbf{v}}_i \\ \mathbf{u}_i \wedge \hat{\mathbf{v}}_i \end{pmatrix}$. This screw annuls all possible reaction wrenches on this axis, $\mathcal{R}_i^T \mathbf{S}_i = 0$. Hence:-

$$\tau_i = \mathcal{W}^T \mathbf{S}_i$$

These dual screws are familiar, though; each $\mathbf{S}_i$ is the instantaneous screw corresponding to rotations about joint i, and in turn we saw in section 6.6 that they were the columns in the robot's jacobian. Hence we may combine all six equations into the following matrix equation:-

$$\mathcal{W}^T \mathbf{J} = (\tau_1, \tau_2, \dots, \tau_6) = \boldsymbol{\tau}^T$$

If the jacobian is non-singular this can be rearranged to give:-

$$\mathcal{W} = (\mathbf{J}^T)^{-1}\boldsymbol{\tau}$$

This result is extremely useful since it allows us to transfer joint torques to the end-effector, hence it is important for force control. It can also be derived using the principle of virtual work.

8.4 Compliance

All robot joints and links are slightly compliant. However, the joint compliance is usually much greater than the link compliance. This springiness in the joints is mainly due to the springiness of gears and shafts but there is also some contribution from the control system and the electric motors themselves. In the simplest model for these processes a deflection $\delta\theta_i$ in the ith joint will produce a torque $\tau_i = k_i\delta\theta_i$, where k_i is the **joint stiffness**; compare this with Hook's law from elementary mechanics.

For a serial manipulator this can be written in the matrix form:-

$$\boldsymbol{\tau} = \mathbf{K}\,\Delta\boldsymbol{\theta}$$

where $\mathbf{K}$ is the diagonal matrix of joint stiffness and is called the stiffness matrix of the robot.

For a six joint manipulator in equilibrium the wrench at the end-effector $\mathcal{W}$, resulting from torques at the joints, is given as we saw in the last section by:-

$$\mathcal{W} = (\mathbf{J}^T)^{-1}\boldsymbol{\tau}$$

with $\mathbf{J}$ the manipulator jacobian. Also the displacement of the end-effector in terms of the joint displacements is:-

$$\Delta\mathbf{x} = \mathbf{J}\,\Delta\boldsymbol{\theta}$$

Putting all this together we get:-

$$\mathcal{W} = (\mathbf{J}^T)^{-1}\mathbf{K}\,\mathbf{J}^{-1}\Delta\mathbf{x}$$

This expresses the wrench caused by a small deflection in the position of the end-effector. Often it is convenient to use the **compliance matrix C** of the manipulator; this is defined as $\mathbf{C} = \mathbf{J}\,\mathbf{K}^{-1}\,\mathbf{J}^T$, so that:-

$$\mathcal{W} = \mathbf{C}^{-1}\Delta\mathbf{x}$$

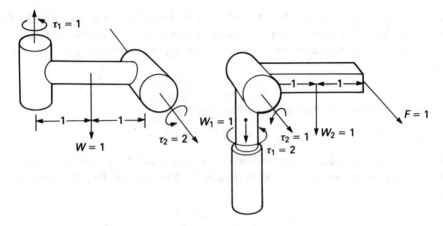

Figure 8.5 (a) A Single Link   (b) Two Robot Links

If the jacobian is singular the compliance matrix is undefined; this corresponds to the arm being infinitely stiff in at least one direction.

In this chapter we have studied problems of static forces and torques. This is a necessary prerequisite to the study of dynamics that we undertake in the next chapter. Our main tool has been the six dimensional vectors called wrenches. These are similar to, but actually dual to, the instantaneous screws that we met in chapter 6. The study of statics has several direct applications in robotics, in particular for gripping objects and for studying the compliance of manipulators.

Exercises

8.1   Consider the rigid body shown in fig. 8.5(a). Calculate the total wrench acting on the body, when the forces and torques are as shown in the diagram. Also find the pitch and axis of this total wrench.

8.2   Two robot links are illustrated in fig. 8.5(b). If both links are in equilibrium with the forces and torques shown, find the reaction wrenches at the joints.

8.3   The Stewart platform is illustrated in fig. 3.6. Assume the centres of the spherical passive joints are located at:-

$$\mathbf{p}_1 = (0,0,0), \qquad \mathbf{p}_2 = (2,0,0), \qquad \mathbf{p}_3 = (1, \sqrt{3}, 0)$$

for the base link, and:-

$$\mathbf{p}_a = (0, 2/\sqrt{3}, 1), \qquad \mathbf{p}_b = (1, -1/\sqrt{3}, 1), \qquad \mathbf{p}_c = (2, 2/\sqrt{3}, 1)$$

for the movable link.

(i) Suppose the magnitude of the forces on the prismatic joints, in suitable units, are:-

$$F_{1a} = 1, \quad F_{1b} = -1, \quad F_{2b} = 1, \quad F_{2c} = -1 \quad F_{3c} = 1, \quad F_{3a} = -1$$

where $F_{i\mu}$ is the magnitude of the force given by the prismatic joint between the points $\mathbf{p}_i$ and $\mathbf{p}_\mu$. Calculate the total wrench acting on the movable link, and find also its pitch and axis.

(ii) It is required that the wrench acting on the movable link must be:-

$$\mathcal{W} = \begin{pmatrix} 6/\sqrt{7} \\ -6\sqrt{3}/\sqrt{7} \\ 0 \\ 0 \\ 2/\sqrt{7} \\ 6\sqrt{3}/\sqrt{7} \end{pmatrix}$$

Calculate the magnitude of the forces that must be applied to the prismatic joint to achieve this.

8.4 Consider a general parallel manipulator with six prismatic joints connecting the base link to the movable link; each with a passive spherical joint at either end. Unit forces along the prismatic joints are given by the wrenches, $\mathcal{F}_1, \mathcal{F}_2, \ldots, \mathcal{F}_6$. The magnitude of the forces exerted by the prismatic joints are $f_1, f_2, \ldots, f_6$. Suppose that when the movable link undergoes a change in position $\Delta \mathbf{x}$, the corresponding changes to the joint lengths are $\delta l_1, \delta l_2, \ldots, \delta l_6$. By considering the work done show that the rows of the manipulator jacobian are given by the transposes of the unit wrenches $\mathcal{F}_i$. (Recall that for a parallel manipulator the jacobian is given by $\Delta \mathbf{l} = \mathbf{J} \Delta \mathbf{x}$.)

9 Dynamics

9.1 Newtonian Mechanics

In this chapter we want to study the mechanics of manipulator arms. We only consider the simplest possible model for robots, that is, of several rigid bodies, the links, connected together. We will ignore such effects as friction in the joints, flexibility of the links and joints and the dynamics of the motors and drives themselves. To begin, we will review some of the relevant classical mechanics.

For a single particle the equation of motion is familiar from Newton's laws of mechanics:-

$$\mathbf{F} = m\dot{\mathbf{v}}$$

Here, $\mathbf{F}$ is the force exerted on the particle m, its mass and $\dot{\mathbf{v}} = \mathrm{d}^2\mathbf{x}/\mathrm{d}t^2$ its acceleration. Although this is the most familiar result of Newtonian mechanics it is slightly misleading for us here. Better is the more general:-

$$\mathbf{F} = \frac{\mathrm{d}}{\mathrm{d}t}\mathbf{p}$$

that is applied force is equal to rate of change of momentum $\mathbf{p}$. The two equations are the same when the mass does not change since the momentum is given by:-

$$\mathbf{p} = m\mathbf{v}$$

Linear momentum is mass times velocity.

The angular momentum $\mathbf{J}$ of the particle is defined as:

$$\mathbf{J} = \mathbf{x} \wedge \mathbf{p}$$

So its rate of change is given by:-

$$\frac{\mathrm{d}}{\mathrm{d}t}\mathbf{J} = \frac{\mathrm{d}}{\mathrm{d}t}(\mathbf{x} \wedge \mathbf{p}) = \mathbf{x} \wedge \mathbf{F}$$

Since $\mathrm{d}\mathbf{x}/\mathrm{d}t \wedge \mathbf{p} = m\mathrm{d}\mathbf{x}/\mathrm{d}t \wedge \mathrm{d}\mathbf{x}/\mathrm{d}t = 0$. This means that the moment of the applied force equals the rate of change of angular momentum.

A rigid body is considered as a collection of particles, held together by internal forces. Luckily, the effects of these internal forces always cancel so they can be ignored. The total mass M of the body is the sum of the masses of the particles. Usually, the particles composing the body are taken to be so small that the total mass can be calculated from the volume integral:-

$$M = \int \rho \, \mathrm{dvol}$$

The density of the body here is ρ.

The centre of mass of the body is a special point in the body where the weight seems to act. The components of its position vector $\mathbf{c}$ are given by the integrals:-

$$\mathbf{c} = \frac{1}{M} \int \mathbf{x} \rho \, \mathrm{dvol}$$

Now the total linear momentum, $\mathbf{p}_{\text{tot}}$, is given by the integrals:-

$$\mathbf{p}_{\text{tot}} = \int \rho \frac{\mathrm{d}\mathbf{x}}{\mathrm{d}t} \, \mathrm{dvol} = \frac{\mathrm{d}}{\mathrm{d}t} \int \rho \mathbf{x} \, \mathrm{dvol} = M \frac{\mathrm{d}\mathbf{c}}{\mathrm{d}t}$$

if we assume the density and hence the mass of the body to be uniform. As far as the forces on the body are concerned, the body can be replaced by a single particle of mass M located at the centre of mass, since we have:-

$$F_{\text{tot}} = M \frac{\mathrm{d}^2 \mathbf{c}}{\mathrm{d}t^2}$$

by differentiating the total momentum.

At any moment the body will be undergoing an instantaneous screw motion; suppose the six dimensional vector of this screw is $\begin{pmatrix} \boldsymbol{\omega} \\ \mathbf{s} \end{pmatrix}$. As we saw in section 6.6, the velocity of any point in the body is given by $\mathrm{d}\mathbf{x}/\mathrm{d}t = \boldsymbol{\omega} \wedge \mathbf{x} + \mathbf{s}$. If we apply this to the centre of mass, then the total linear momentum is given by:-

$$\mathbf{p}_{\text{tot}} = M \boldsymbol{\omega} \wedge \mathbf{c} + M \mathbf{s}$$

In a similar fashion we can write down the total angular momentum as the volume integral:-

$$J_{\text{tot}} = \int \rho \mathbf{x} \wedge \frac{\mathrm{d}\mathbf{x}}{\mathrm{d}t} \, \mathrm{dvol}$$

Now, if we use the fact that the body is performing an instantaneous screw, we may replace $\mathrm{d}\mathbf{x}/\mathrm{d}t$ with $\boldsymbol{\omega} \wedge \mathbf{x} + \mathbf{s}$. The total angular momentum can now be written as:-

$$J_{\text{tot}} = \int \{\rho \mathbf{x} \wedge (\boldsymbol{\omega} \wedge \mathbf{x}) + \rho \mathbf{x} \wedge \mathbf{s}\} \, \mathrm{dvol} = \int \rho \mathbf{x} \wedge (\boldsymbol{\omega} \wedge \mathbf{x}) \, \mathrm{dvol} + M \mathbf{c} \wedge \mathbf{s}$$

The integral term we are left with contains the vector triple product $\mathbf{x} \wedge (\boldsymbol{\omega} \wedge \mathbf{x})$. If we let:-

$$\mathbf{x} = \begin{pmatrix} x \\ y \\ z \end{pmatrix} \qquad \text{and} \qquad \boldsymbol{\omega} = \begin{pmatrix} \omega_x \\ \omega_y \\ \omega_z \end{pmatrix}$$

then the triple product can be computed:-

$$\mathbf{x} \wedge (\boldsymbol{\omega} \wedge \mathbf{x}) = \begin{pmatrix} (y^2 + z^2)\omega_x - xy\omega_y - xz\omega_z \\ -xy\omega_x + (x^2 + z^2)\omega_y - yz\omega_z \\ -xz\omega_x - yz\omega_y + (x^2 + y^2)\omega_z \end{pmatrix}$$

Since the components of $\boldsymbol{\omega}$ only appear linearly in these expressions we may write the result in a matrix form:-

$$\mathbf{x} \wedge (\boldsymbol{\omega} \wedge \mathbf{x}) = \begin{pmatrix} (y^2 + z^2) & -xy & -xz \\ -xy & (x^2 + z^2) & -yz \\ -xz & -yz & (x^2 + y^2) \end{pmatrix} \begin{pmatrix} \omega_x \\ \omega_y \\ \omega_z \end{pmatrix}$$

This means that the total angular momentum can be written as:-

$$J_{\text{tot}} = \mathbf{I}\boldsymbol{\omega} + M\mathbf{c} \wedge \mathbf{s}$$

The matrix $\mathbf{I}$ is called the 3×3 **inertia matrix** of the body: its components only depend on the mass distribution of the body, and they are given by:-

$$\mathbf{I} = \begin{pmatrix} \int \rho(y^2 + z^2)\,\text{dvol} & -\int \rho xy\,\text{dvol} & -\int \rho xz\,\text{dvol} \\ -\int \rho xy\,\text{dvol} & \int \rho(x^2 + z^2)\,\text{dvol} & -\int \rho yz\,\text{dvol} \\ -\int \rho xz\,\text{dvol} & -\int \rho yz\,\text{dvol} & \int \rho(x^2 + y^2)\,\text{dvol} \end{pmatrix}$$

From the above we have two three-vector equations of motion for a rigid body:-

$$\tau_{\text{tot}} = \frac{d}{dt}(\mathbf{I}\boldsymbol{\omega} + M\mathbf{c} \wedge \mathbf{s})$$

$$F_{\text{tot}} = \frac{d}{dt}(M\boldsymbol{\omega} \wedge \mathbf{c} + M\mathbf{s})$$

In robotics we are fortunate in that we usually do not have to solve these differential equations. More often the desired motion of the rigid body is known and it is the total forces and torques needed to produce this motion that must be computed. If, as is normal, the mass of the rigid body is constant, then the equations for the total force can be tidied up a little; however, the inertia matrix changes as the body moves.

In chapters 6 and 8 we developed a six-vector notation for representing screws and wrenches. The equations of motion can be written very compactly as a single six-vector equation:-

$$\mathcal{W} = \frac{d}{dt}(\mathbf{N}\mathbf{V})$$

where $\mathcal{W} = \begin{pmatrix} \tau \\ F \end{pmatrix}$ is the total wrench acting on the body, and $\mathbf{V} = \begin{pmatrix} \boldsymbol{\omega} \\ \mathbf{s} \end{pmatrix}$ is the instantaneous screw of the body. The matrix $\mathbf{N}$ is the 6×6 inertia matrix of the body, and its components can be written in partitioned form as:-

$$\mathbf{N} = \begin{pmatrix} \mathbf{I} & M\mathbf{C} \\ M\mathbf{C}^T & M\mathbf{I} \end{pmatrix}$$

Here, $\mathbf{C}$ is the antisymmetric matrix satisfying $\mathbf{C}\mathbf{u} = \mathbf{c} \wedge \mathbf{u}$ for any vector $\mathbf{u}$, see exercise 2.7. And $\mathbf{I}$ is just the 3×3 identity matrix. Notice that $\mathbf{N}$ is in fact a symmetric matrix.

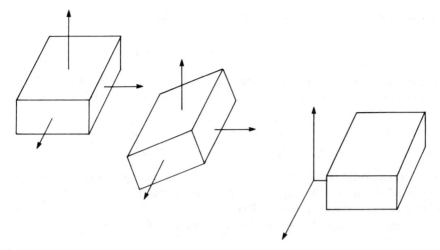

Figure 9.1 A Rectangular Block in Three Positions

In a real control system for a robot this compact notation would not be of too much help since all six components would have to be computed separately. But for the algebraic manipulations that we must perform in the following, the neater the notation the better.

9.2 Moments of Inertia

We will restrict ourselves to a single example for the calculation of the inertia matrix. The subject of calculating these integrals is well covered in many textbooks on mechanics and applied mathematics. Consider the rectangular block shown in fig. 9.1. Suppose the block has sides of length $2a$, $2b$ and $2c$. In the first instance we will assume that the sides are aligned with the x, y and z axes respectively and that the origin is located at the centre of the block. If the block has uniform density ρ, its total mass will be:-

$$M = 8\rho abc$$

The first term in the inertia matrix is:-

$$\int \rho(y^2 + z^2)\,\mathrm{dvol} = \rho \int_{-a}^{+a} \mathrm{d}x \left(\int_{-b}^{+b} y^2 \;\mathrm{d}y \int_{-c}^{+c} \mathrm{d}z + \int_{-b}^{+b} \mathrm{d}y \int_{-c}^{+c} z^2\,\mathrm{d}z \right)$$

This is easily computed:-

$$\int \rho(y^2 + z^2)\,\mathrm{dvol} = 8\rho a\big(\frac{b^3}{3}c + b\frac{c^3}{3}\big) = \frac{M}{3}(b^2 + c^2)$$

Clearly the other diagonal terms are very similar and can be found simply by relabelling the variables.

The first off-diagonal element in the matrix is given by:-

$$\int \rho xy \, \mathrm{dvol} = \rho \int_{-a}^{+a} x \, \mathrm{d}x \int_{-b}^{+b} y \, \mathrm{d}y \int_{-c}^{+c} \mathrm{d}z$$

The first two integrals in the product both vanish. Again by relabelling the variables it can be seen that all off-diagonal entries are zero. The 3×3 inertia matrix is thus given by:-

$$\mathbf{I} = \begin{pmatrix} \frac{M}{3}(b^2 + c^2) & 0 & 0 \\ 0 & \frac{M}{3}(a^2 + c^2) & 0 \\ 0 & 0 & \frac{M}{3}(a^2 + b^2) \end{pmatrix}$$

The inertia matrix takes this simple form because we have chosen the origin and axes to be in a particularly 'nice' configuration. What would happen if we were to rotate the block about the x-axis through an angle θ? Rather than compute a lot of integrals again we will derive a general theorem. Recall that the inertia matrix is defined by:-

$$\mathbf{I}\boldsymbol{\omega} = \int \rho \mathbf{x} \wedge (\boldsymbol{\omega} \wedge \mathbf{x}) \, \mathrm{dvol}$$

Now if we rotate the body using a rotation $\mathbf{R}$, each point $\mathbf{x}$ turns into $\mathbf{R}\mathbf{x}$ so the new inertia matrix satisfies:-

$$\mathbf{I}'\boldsymbol{\omega} = \int \rho \mathbf{R}\mathbf{x} \wedge (\boldsymbol{\omega} \wedge \mathbf{R}\mathbf{x}) \, \mathrm{dvol}$$

Since for any three-vectors we have $\mathbf{R}(\mathbf{a} \wedge \mathbf{b}) = \mathbf{R}\mathbf{a} \wedge \mathbf{R}\mathbf{b}$, we can rewrite the integral as:-

$$\mathbf{I}'\boldsymbol{\omega} = \mathbf{R} \int \rho \mathbf{x} \wedge (\mathbf{R}^{-1}\boldsymbol{\omega} \wedge \mathbf{x}) \, \mathrm{dvol} = \mathbf{R}\mathbf{N}\mathbf{R}^{-1}\boldsymbol{\omega}$$

Hence, the new inertia matrix is given by the similarity transformation:-

$$\mathbf{I}' = \mathbf{R}\,\mathbf{N}\,\mathbf{R}^{-1}$$

This result is sometimes referred to as the tensor property of the inertia matrix. ΓΓΓmmber that rotation matrices are orthogonal, so $\mathbf{R}^{-1} = \mathbf{R}^T$.

For the example mentioned above, the rotation matrix is:-

$$\mathbf{R} = \begin{pmatrix} 1 & 0 & 0 \\ 0 & \cos\theta & -\sin\theta \\ 0 & \sin\theta & \cos\theta \end{pmatrix}$$

and so the inertia matrix in this position is:-

$$\mathbf{I}' =$$
$$\begin{pmatrix} 1 & 0 & 0 \\ 0 & \cos\theta & -\sin\theta \\ 0 & \sin\theta & \cos\theta \end{pmatrix} \begin{pmatrix} \frac{M}{3}(b^2 + c^2) & 0 & 0 \\ 0 & \frac{M}{3}(a^2 + c^2) & 0 \\ 0 & 0 & \frac{M}{3}(a^2 + b^2) \end{pmatrix} \begin{pmatrix} 1 & 0 & 0 \\ 0 & \cos\theta & \sin\theta \\ 0 & -\sin\theta & \cos\theta \end{pmatrix}$$

$$= \frac{M}{3} \begin{pmatrix} b^2 + c^2 & 0 & 0 \\ 0 & a^2 + b^2\sin^2\theta + c^2\cos^2\theta & (c^2 - b^2)\cos\theta\sin\theta \\ 0 & (c^2 - b^2)\cos\theta\sin\theta & a^2 + b^2\cos^2\theta + c^2\sin^2\theta \end{pmatrix}$$

Next we look at what happens to the inertia matrix if the rigid body undergoes a translation. Once again we start with the definition of the inertia matrix:-

$$\mathbf{I}\boldsymbol{\omega} = \int \rho \mathbf{x} \wedge (\boldsymbol{\omega} \wedge \mathbf{x}) \quad \text{dvol}$$

After a translation by $\mathbf{t}$, the points $\mathbf{x}$ become $\mathbf{x} + \mathbf{t}$. So the new inertia matrix satisfies:-

$$\mathbf{I}''\boldsymbol{\omega} = \int \rho(\mathbf{x} + \mathbf{t}) \wedge (\boldsymbol{\omega} \wedge (\mathbf{x} + \mathbf{t})) \,\text{dvol}$$

Expanding the triple product gives:-

$$(\mathbf{x} + \mathbf{t}) \wedge (\boldsymbol{\omega} \wedge (\mathbf{x} + \mathbf{t})) = \mathbf{x} \wedge (\boldsymbol{\omega} \wedge \mathbf{x}) + \mathbf{t} \wedge (\boldsymbol{\omega} \wedge \mathbf{x}) + \mathbf{x} \wedge (\boldsymbol{\omega} \wedge \mathbf{t}) + \mathbf{t} \wedge (\boldsymbol{\omega} \wedge \mathbf{t})$$

Since $\mathbf{t}$ and $\boldsymbol{\omega}$ are fixed, and remembering that:-

$$\int \rho \mathbf{x} \,\text{dvol} = M\mathbf{c}$$

performing the integrals gives:-

$$\mathbf{I}''\boldsymbol{\omega} = \mathbf{I}\boldsymbol{\omega} + M\mathbf{t} \wedge (\boldsymbol{\omega} \wedge \mathbf{c}) + M\mathbf{c} \wedge (\boldsymbol{\omega} \wedge \mathbf{t}) + M\mathbf{t} \wedge (\boldsymbol{\omega} \wedge \mathbf{t})$$

where $\mathbf{c}$ is the position vector of the body's centre of mass in the original position. Using the fact that we can use an antisymmetric matrix to represent the effect of taking a vector product with a fixed vector, we may write the result as:-

$$\mathbf{I}''\boldsymbol{\omega} = (\mathbf{I} - M\mathbf{T}\,\mathbf{C} - M\mathbf{C}\,\mathbf{T} - M\mathbf{T}^2)\boldsymbol{\omega}$$

Since $\boldsymbol{\omega}$ is arbitrary here, the new inertia matrix can be written:-

$$\mathbf{I}'' = \mathbf{I} - M\mathbf{T}\,\mathbf{C} - M\mathbf{C}\,\mathbf{T} - M\mathbf{T}^2$$

This result is sometimes called the parallel axis theorem. As an example of its use, consider translating the block of fig. 9.1 by an arbitrary amount, that is:-

$$\mathbf{t} = \begin{pmatrix} t_x \\ t_y \\ t_z \end{pmatrix} \quad \text{and hence} \quad \mathbf{T} = \begin{pmatrix} 0 & -t_z & t_y \\ t_z & 0 & -t_x \\ -t_y & t_x & 0 \end{pmatrix}$$

In this case, the body is originally positioned so that its centre of mass coincides with the origin, therefore $\mathbf{C} = 0$. The new inertia matrix is thus:-

$$
\begin{aligned}
\mathbf{I}'' &= \begin{pmatrix} \frac{M}{3}(b^2 + c^2) & 0 & 0 \\ 0 & \frac{M}{3}(a^2 + c^2) & 0 \\ 0 & 0 & \frac{M}{3}(a^2 + b^2) \end{pmatrix} \\
&\quad -M \begin{pmatrix} -(t_y^2 + t_z^2) & t_x t_y & t_x t_z \\ t_x t_y & -(t_x^2 + t_z^2) & t_y t_z \\ t_x t_z & t_y t_z & -(t_x^2 + t_y^2) \end{pmatrix} \\
&= M \begin{pmatrix} \frac{1}{3}(b^2 + c^2) + (t_y^2 + t_z^2) & -t_x t_y & -t_x t_z \\ -t_x t_y & \frac{1}{3}(a^2 + c^2) + (t_x^2 + t_y^2) & -t_y t_z \\ -t_x t_z & -t_y t_z & \frac{1}{3}(a^2 + b^2) + (t_x^2 + t_y^2) \end{pmatrix}
\end{aligned}
$$

If we are prepared to use the six-vector notation introduced above for wrenches and screws, then the two previous results can be combined into a single neat formula. In exercise 6.6 we found that under a rigid body motion a screw S transforms to HS, where:-

$$S = \begin{pmatrix} \omega \\ s \end{pmatrix} \quad \text{and} \quad H = \begin{pmatrix} R & 0 \\ TR & R \end{pmatrix}$$

Actually in exercise 6.6 we only saw this for screws of the form $S = \begin{pmatrix} \hat{v} \\ u \wedge \hat{v} \end{pmatrix}$, but it is not hard to show that the above generalization is true.

Under the same rotation and translation the 6×6 inertia matrix N will be transformed into:-

$$N' = (H^T)^{-1} N H^{-1}$$

Unlike the rotation matrices the matrix H is not orthogonal, in fact:-

$$(H^T)^{-1} = \begin{pmatrix} R & TR \\ 0 & R \end{pmatrix} \quad \text{since} \quad H^{-1} = \begin{pmatrix} R^T & 0 \\ -R^T T & R^T \end{pmatrix}$$

We recover the original relations by multiplying out the three partitioned matrices. However, it must be remembered that this relation is for a rotation followed by a translation whereas our original second result was for a translation on its own. The relation also gives us that the new centre of mass becomes:-

$$C' = R C R^T + T$$

9.3   Time Derivatives of the Inertia

In the case of a particle with constant mass the linear momentum is easy to differentiate and so the equation of motion for the particle becomes $F = m\dot{v}$. Here, we want to do the same thing for a rigid body, that is we will assume the body has constant mass and shape. To begin let us look at the case of pure rotations. At $t = 0$ the rigid body has a 3×3 inertia matrix which we will denote I_0. At subsequent times the body will have undergone a rotation about the origin and so its inertia matrix will be given by:-

$$I = R I_0 R^T$$

where the rotation matrix R will be a function of time. Differentiating this with respect to time simply gives:-

$$\frac{d}{dt}I = \dot{R} I_0 R^T + R I_0 \dot{R}^T$$

So we can write the time derivative of the angular momentum, $J = I\omega$, as:-

$$\frac{d}{dt}J = I\dot{\omega} + \dot{R} I_0 R^T \omega + R I_0 \dot{R}^T \omega$$

Now as we saw in section 6.5, when $t = 0$ the matrix $\dot{\mathbf{R}}(0)$ has the same effect as the vector product with the angular velocity $\boldsymbol{\omega}$:-

$$\dot{\mathbf{R}}(0)\mathbf{u} = \boldsymbol{\omega} \wedge \mathbf{u}$$

for any vector $\mathbf{u}$. At $t = 0$ the derivative of the angular momentum is therefore:-

$$\frac{d}{dt}\mathbf{J}(0) = \mathbf{I}_0\dot{\boldsymbol{\omega}} + \boldsymbol{\omega} \wedge (\mathbf{I}_0\boldsymbol{\omega})$$

since $\dot{\mathbf{R}}^T\boldsymbol{\omega} = -\boldsymbol{\omega} \wedge \boldsymbol{\omega} = 0$. Using the same arguments as in section 6.5, there is nothing special about the time $t = 0$, we could have chosen to begin measuring time at any instant, so the above equation must be generally true:-

$$\frac{d}{dt}\mathbf{J} = \mathbf{I}\dot{\boldsymbol{\omega}} + \boldsymbol{\omega} \wedge (\mathbf{I}\boldsymbol{\omega})$$

This is now equal to the torque applied to the rigid body, and the result is known as **Euler's equation**:-

$$\boldsymbol{\tau} = \mathbf{I}\dot{\boldsymbol{\omega}} + \boldsymbol{\omega} \wedge (\mathbf{I}\boldsymbol{\omega})$$

Actually this is not quite true: Euler's equation is usually written in terms of co-ordinates at rest with respect to the body; in such co-ordinates the inertia matrix is constant. We are only using a single, fixed co-ordinate frame, so for us the inertia matrix varies; however, the form of the equations is exactly the same.

For a general rigid body motion we can mimic the derivation just given but with the 6×6 matrices. First we will need to know about the derivatives of the matrix:-

$$\mathbf{H} = \begin{pmatrix} \mathbf{R} & 0 \\ \mathbf{TR} & \mathbf{R} \end{pmatrix}$$

this is simply:-

$$\frac{d}{dt}\mathbf{H} = \begin{pmatrix} \dot{\mathbf{R}} & 0 \\ \dot{\mathbf{T}}\mathbf{R} + \mathbf{T}\dot{\mathbf{R}} & \dot{\mathbf{R}} \end{pmatrix}$$

We have seen that $\dot{\mathbf{R}}(0)\mathbf{u} = \boldsymbol{\omega} \wedge \mathbf{u}$, for any vector $\mathbf{u}$. Since the translation part of the rigid transformation is $\mathbf{T}\mathbf{u} = \mathbf{t} \wedge \mathbf{u}$, we must have $\dot{\mathbf{T}}(0)\mathbf{u} = \mathbf{s} \wedge \mathbf{u}$. For arbitrary $\mathbf{u}$, and where $\mathbf{s} = \dot{\mathbf{t}}$ is the linear velocity component of the body's instantaneous screw, see section 6.6. Now when $t = 0$ we have $\mathbf{R} = \mathbf{I}$ and $\mathbf{T} = 0$. Hence, for an arbitrary instantaneous screw $(\boldsymbol{\psi}, \mathbf{u})^T$, we can write:-

$$\frac{d}{dt}\mathbf{H}(0)\begin{pmatrix} \boldsymbol{\psi} \\ \mathbf{u} \end{pmatrix} = \begin{pmatrix} \dot{\mathbf{R}}(0) & 0 \\ \dot{\mathbf{T}}(0) & \dot{\mathbf{R}}(0) \end{pmatrix}\begin{pmatrix} \boldsymbol{\psi} \\ \mathbf{u} \end{pmatrix} = \begin{pmatrix} \boldsymbol{\omega} \wedge \boldsymbol{\psi} \\ \mathbf{s} \wedge \boldsymbol{\psi} + \boldsymbol{\omega} \wedge \mathbf{u} \end{pmatrix}$$

Again this result does not depend on where we begin measuring time, so we can write:-

$$\frac{d}{dt}\mathbf{H}\begin{pmatrix} \boldsymbol{\psi} \\ \mathbf{u} \end{pmatrix} = \begin{pmatrix} \boldsymbol{\omega} \wedge \boldsymbol{\psi} \\ \mathbf{s} \wedge \boldsymbol{\psi} + \boldsymbol{\omega} \wedge \mathbf{u} \end{pmatrix}$$

This is further simplified if we adopt the following notation; for two six-vectors we will write:-

$$\begin{pmatrix} \psi_1 \\ \mathbf{u_1} \end{pmatrix} \wedge \begin{pmatrix} \psi_2 \\ \mathbf{u_2} \end{pmatrix} = \begin{pmatrix} \psi_1 \wedge \psi_2 \\ \mathbf{u_1} \wedge \psi_2 + \psi_1 \wedge \mathbf{u_2} \end{pmatrix}$$

This new operation is the vector product of two instantaneous screws. Although we use the symbol '$\wedge$' for the vector products of both three and six component vectors, no confusion should arise. So now we can write the time derivative of an arbitrary rigid motion as:-

$$\frac{d}{dt}\mathbf{H}\begin{pmatrix} \psi \\ \mathbf{u} \end{pmatrix} = \begin{pmatrix} \omega \\ \mathbf{s} \end{pmatrix} \wedge \begin{pmatrix} \psi \\ \mathbf{u} \end{pmatrix}$$

By similar arguments we find that the time derivative of the matrix $(\mathbf{H}^T)^{-1}$ acting on an arbitrary wrench:-

$$\frac{d}{dt}(\mathbf{H}^T)^{-1}\begin{pmatrix} \gamma \\ \mathbf{E} \end{pmatrix} = \begin{pmatrix} \omega \wedge \gamma + \mathbf{s} \wedge \mathbf{E} \\ \omega \wedge \gamma \end{pmatrix}$$

This can be written using another new operation, which will be denoted by curly braces:-

$$\frac{d}{dt}(\mathbf{H}^T)^{-1}\begin{pmatrix} \gamma \\ \mathbf{E} \end{pmatrix} = \left\{ \begin{pmatrix} \omega \\ \mathbf{s} \end{pmatrix}, \begin{pmatrix} \gamma \\ \mathbf{E} \end{pmatrix} \right\}$$

This is a new type of operation: we are 'multiplying' a wrench by a screw. The result is another wrench:-

$$\left\{ \begin{pmatrix} \psi \\ \mathbf{u} \end{pmatrix}, \begin{pmatrix} \gamma \\ \mathbf{E} \end{pmatrix} \right\} = \begin{pmatrix} \psi \wedge \gamma + \mathbf{u} \wedge \mathbf{E} \\ \psi \wedge \mathbf{E} \end{pmatrix}$$

At last we are able to write down the six-vector equations of motion:-

$$W = \frac{d}{dt}(\mathbf{N}\,\mathbf{V}) = \mathbf{N}\,\dot{\mathbf{V}} + \left\{ \mathbf{V}, \mathbf{N}\,\mathbf{V} \right\} - \mathbf{N}(\mathbf{V} \wedge \mathbf{V})$$

The vector product of an instantaneous screw with itself is zero so the equations reduce to:-

$$W = \mathbf{N}\,\dot{\mathbf{V}} + \left\{ \mathbf{V}, \mathbf{N}\,\mathbf{V} \right\}$$

These equations summarize both Newton's and Euler's equations and are thus called the Newton-Euler equations.

It is more usual to use separate equations for the motion of the centre of mass and the motion about the centre of mass; that is, Euler's equation for the torque and Newton's for the force. The general form of these can be found by looking at the above equation in partitioned form:-

$$\begin{pmatrix} \tau_{\text{tot}} \\ \mathbf{F}_{\text{tot}} \end{pmatrix} = \begin{pmatrix} \mathbf{I} & M\mathbf{C} \\ M\mathbf{C}^T & M\mathbf{I} \end{pmatrix} \begin{pmatrix} \dot{\omega} \\ \dot{\mathbf{s}} \end{pmatrix} + \left\{ \begin{pmatrix} \omega \\ \mathbf{s} \end{pmatrix}, \begin{pmatrix} \mathbf{I} & M\mathbf{C} \\ M\mathbf{C}^T & M\mathbf{I} \end{pmatrix} \begin{pmatrix} \omega \\ \mathbf{s} \end{pmatrix} \right\}$$

Which, after a little simplification, becomes:-

$$\begin{aligned} \tau_{\text{tot}} &= \mathbf{I}\dot{\omega} + \omega \wedge (\mathbf{I}\omega) + M\mathbf{c} \wedge (\dot{\mathbf{s}} + \omega \wedge \mathbf{s}) \\ \mathbf{F}_{\text{tot}} &= M(\dot{\mathbf{s}} + (\dot{\omega} - \mathbf{s}) \wedge \mathbf{c}) \end{aligned}$$

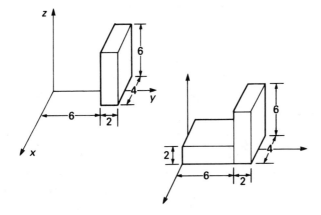

Figure 9.2 Single Block (a)   and Two blocks Joined (b)

Exercises

9.1 (i) Calculate the 3×3 inertia matrix of a rectangular block of mass 1 unit and sides 2, 4 and 6 units, positioned as shown in fig. 9.2(a).

(ii) Show that the 3×3 inertia matrix of two rigid bodies rigidly joined together is the sum of their individual inertia matrices. Hence find the inertia matrix for the body consisting of two rectangular blocks as shown in fig. 9.2(b).

9.2 (i) Show that for any rigid body there is a co-ordinate frame in which the inertia matrix takes the form:-

$$\mathbf{N} = \begin{pmatrix} \mathbf{D} & 0 \\ 0 & M\mathbf{I} \end{pmatrix}$$

where $\mathbf{D}$ is a diagonal matrix with positive entries along the leading diagonal.

(ii) Is it possible, for any rigid body, to find a rectangular block with the same mass and inertia matrix?

9.3 (i) Show that for any instantaneous screw $\mathbf{S}$:-

$$\mathbf{S} \wedge \mathbf{S} = 0$$

(ii) Given a wrench $\mathcal{W}$ and two screws $\mathbf{S}_1$ and $\mathbf{S}_2$ show that:-

$$\left\{ \mathbf{S}_1, \mathcal{W} \right\}^T \mathbf{S}_2 = \mathcal{W}^T (\mathbf{S}_1 \wedge \mathbf{S}_2)$$

9.4 One Link

In order to see how the rigid body mechanics described above can be used, we will look at the problem of finding the equations of motion for a single driven link. This simple

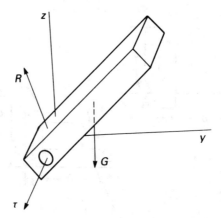

Figure 9.3 A Single Driven Link

example can be approached in many ways and does not really need the powerful techniques we have developed. However, the following is only intended as an introduction to the more complicated situation of several rigid bodies jointed together.

To begin, as with all mechanics problems, we must draw a diagram of the forces and torques involved, see fig. 9.3. Although we are assuming that the joint is a revolute joint, our techniques work equally well for prismatic and helical joints.

The total wrench acting on the body has three sources: the torque τ, due to the motor, the weight of the link $\mathcal{G}$ and the reaction wrench $\mathcal{R}$ at the bearing. The total wrench can be written:-

$$\mathcal{W} = \tau + \mathcal{G} + \mathcal{R}$$

With the co-ordinates as shown we can write the torque as:-

$$\tau = \begin{pmatrix} \tau\mathbf{i} \\ 0 \end{pmatrix}$$

with τ the magnitude of the torque. The weight can also be written as:-

$$\mathcal{G} = \begin{pmatrix} -Mg\mathbf{c} \wedge \mathbf{k} \\ -Mg\mathbf{k} \end{pmatrix}$$

Here g is the acceleration due to gravity and M the mass of the link. As we saw above the equation of motion for a rigid body is:-

$$\mathcal{W} = \mathbf{N}\dot{\mathbf{V}} + \left\{ \mathbf{V}, \mathbf{N}\mathbf{V} \right\}$$

The difficulty is that we do not know what the reaction wrench is. All we know about the reaction is that it will be annulled by the joint screw $\mathbf{S} = \begin{pmatrix} \mathbf{i} \\ 0 \end{pmatrix}$; see section 8.3. So we can pair the Newton-Euler equation with this instantaneous screw to get rid of the reaction wrench:-

$$\mathcal{W}^T \mathbf{S} = \boldsymbol{\tau} - Mg\mathbf{c} \wedge \mathbf{k} \cdot \mathbf{i} = \boldsymbol{\tau} - Mg\mathbf{j} \cdot \mathbf{c}$$

Now, since the body is constrained to rotate about the joint, the velocity screw of the body will simply be:-

$$\mathbf{V} = \mathbf{S}\dot{\theta}$$

with θ, the joint angle. So the right-hand side of the Newton-Euler equations become:-

$$\mathcal{W}^T \mathbf{S} = \mathbf{S}^T \mathbf{N} \mathbf{S} \ddot{\theta} + \left\{ \mathbf{S}, \mathbf{N}\mathbf{S} \right\}^T \mathbf{S} \dot{\theta}^2$$

From the results of exercise 9.3, we can see that the second term vanishes. The Newton-Euler equation can now be written:-

$$\boldsymbol{\tau} = \mathbf{i} \cdot (\mathbf{Ii}) + Mg\mathbf{j} \cdot \mathbf{c}$$

Finally we must take into account the fact that the inertia matrix and the centre of mass are position dependent. In the initial position, where $\theta = 0$, the inertia will be $\mathbf{I}(0)$ and the centre of mass $\mathbf{c}(0)$. For simplicity let us assume that $\mathbf{c}(0)$ only has components in the y-direction. When the body has undergone a rotation of θ radians the new inertia matrix and centre of mass will be:-

$$\mathbf{I}(\theta) = \mathbf{R}\,\mathbf{I}(0)\mathbf{R}^T \qquad \text{and} \qquad \mathbf{c}(\theta) = \mathbf{R}\,\mathbf{c}(0);$$

where

$$\mathbf{R} = \begin{pmatrix} 1 & 0 & 0 \\ 0 & \cos\theta & -\sin\theta \\ 0 & \sin\theta & \cos\theta \end{pmatrix}$$

Hence, the equations of motion simplify to:-

$$\boldsymbol{\tau} = I_{11}(0)\ddot{\theta} + Mg\cos\theta\, c_y(0)$$

since $\mathbf{c}(0)$ has no components in the y-direction.

Although this is a rather cumbersome method for the above example it illustrates the general approach. The idea is to write down the Newton-Euler equations of motion for each rigid body, then project out components of the wrenches we are interested in. As another simple example, suppose we were interested in the reaction forces at the joint. If we were designing the bearings at a robot's joint, it would be important to know the loads which the bearings might be subjected to. Let us compute the reaction force in the z-direction, R_z. The Newton-Euler equation is the same as above, but now we must pair with the screw, $\mathbf{P} = \begin{pmatrix} \mathbf{0} \\ \mathbf{k} \end{pmatrix}$:-

$$R_z = \mathbf{S}^T \mathbf{N} \mathbf{P} \ddot{\theta} + \left\{ \mathbf{S}, \mathbf{N}\mathbf{S} \right\}^T \mathbf{P}\dot{\theta}^2 + \mathbf{W}^T \mathbf{P}$$

Since we know how these matrices are partitioned we can simplify the equation:-

$$R_z = -Mc_y\ddot{\theta} + Mc_z\dot{\theta}^2 - Mg$$

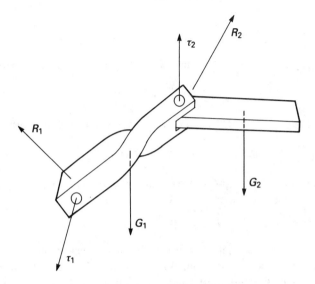

Figure 9.4 Two Driven Links

Once again if we assume that initially the centre of mass lies along the y-axis, the equation becomes:-

$$R_z = -Mc_y(0)\cos\theta\,\ddot{\theta} - Mc_y(0)\sin\theta\,\dot{\theta}^2 - Mg$$

9.5   Two Links

In this section we extend the work already done to two driven links. For the force diagram, see fig. 9.4. First of all since there are now two links, there will be two Newton-Euler equations. They are:-

$$\tau_2 + \mathcal{R}_2 + \mathcal{G}_2 = \mathbf{N}_2\dot{\mathbf{V}}_2 + \left\{\mathbf{V}_2, \mathbf{N}_2\mathbf{V}_2\right\}$$

$$\tau_1 + \mathcal{R}_1 + \mathcal{G}_1 - \tau_2 - \mathcal{R}_2 = \mathbf{N}_1\dot{\mathbf{V}}_1 + \left\{\mathbf{V}_1, \mathbf{N}_1\mathbf{V}_1\right\}$$

In preparation for pairing these equations with the joint screws we can tidy them up a little by adding the first to the second:-

$$\tau_2 + \mathcal{R}_2 = \mathbf{N}_2\dot{\mathbf{V}}_2 + \left\{\mathbf{V}_2, \mathbf{N}_2\mathbf{V}_2\right\} - \mathcal{G}_2$$

$$\tau_1 + \mathcal{R}_1 = \mathbf{N}_1\dot{\mathbf{V}}_1 + \left\{\mathbf{V}_1, \mathbf{N}_1\mathbf{V}_1\right\} + \mathbf{N}_2\dot{\mathbf{V}}_2 + \left\{\mathbf{V}_2, \mathbf{N}_2\mathbf{V}_2\right\} - \mathcal{G}_1 - \mathcal{G}_2$$

Now each equation only involves the torque produced by one motor. Thus, pairing with the joint screws $\mathbf{S}_1$ and $\mathbf{S}_2$ gets rid of the reaction wrenches and leaves the magnitude of the torques:-

$$\tau_2 = \dot{\mathbf{V}}_2^T \mathbf{N}_2 \mathbf{S}_2 + \left\{ \mathbf{V}_2, \mathbf{N}_2 \mathbf{V}_2 \right\}^T \mathbf{S}_2 - \mathcal{G}_2^T \mathbf{S}_2$$
$$\tau_1 = \dot{\mathbf{V}}_1^T \mathbf{N}_1 \mathbf{S}_1 + \left\{ \mathbf{V}_1, \mathbf{N}_1 \mathbf{V}_1 \right\}^T \mathbf{S}_1 + \dot{\mathbf{V}}_2^T \mathbf{N}_2 \mathbf{S}_1 + \left\{ \mathbf{V}_2, \mathbf{N}_2 \mathbf{V}_2 \right\}^T \mathbf{S}_1 - \mathcal{G}_1^T \mathbf{S}_1 - \mathcal{G}_2^T \mathbf{S}_1$$

These are the equations of motion: they can be made a little neater with the aid of exercise 9.3:-

$$\tau_2 = \dot{\mathbf{V}}_2^T \mathbf{N}_2 \mathbf{S}_2 + \mathbf{V}_2^T \mathbf{N}_2 (\mathbf{V}_2 \wedge \mathbf{S}_2) - \mathcal{G}_2^T \mathbf{S}_2$$
$$\tau_1 = (\dot{\mathbf{V}}_1^T \mathbf{N}_1 + \dot{\mathbf{V}}_2^T \mathbf{N}_2) \mathbf{S}_1 + \mathbf{V}_1^T \mathbf{N}_1 (\mathbf{V}_1 \wedge \mathbf{S}_1) + \mathbf{V}_2^T \mathbf{N}_2 (\mathbf{V}_2 \wedge \mathbf{S}_1) - \mathcal{G}_1^T \mathbf{S}_1 - \mathcal{G}_2^T \mathbf{S}_1$$

However, it is more usual to express the equations of motion in terms of the joint angles rather than the $\mathbf{V}$'s which are the velocity screws of the links. The velocity screw of a link is given by $\mathbf{V} = \mathbf{J}\dot{\boldsymbol{\theta}}$, and from section 6.6, we know that the columns of the jacobian $\mathbf{J}$ are the joint screws. Using the appropriate jacobian we have:-

$$\mathbf{V}_1 = \dot{\theta}_1 \mathbf{S}_1 \qquad \text{and} \qquad \mathbf{V}_2 = \dot{\theta}_1 \mathbf{S}_1 + \dot{\theta}_2 \mathbf{S}_2$$

We also need the link acceleration screws, the $\dot{\mathbf{V}}$'s. Since the first joint is fixed we have:-

$$\dot{\mathbf{V}}_1 = \ddot{\theta}_1 \mathbf{S}_1$$

But the second joint moves, so:-

$$\dot{\mathbf{V}}_2 = \ddot{\theta}_1 \mathbf{S}_1 + \ddot{\theta}_2 \mathbf{S}_2 + \dot{\theta}_2 \dot{\mathbf{S}}_2$$

We know how the second joint moves; it simply turns about the first joint:-

$$\mathbf{S}_2(t) = \mathbf{H}_1(t) \mathbf{S}_2(0)$$

So using the results of section 9.3, the derivative of this screw is:-

$$\frac{\mathrm{d}}{\mathrm{d}t} \mathbf{S}_2 = \mathbf{V}_1 \wedge \mathbf{S}_2 = \dot{\theta}_1 \mathbf{S}_1 \wedge \mathbf{S}_2$$

The acceleration of the second link is thus:-

$$\dot{\mathbf{V}}_2 = \ddot{\theta}_1 \mathbf{S}_1 + \ddot{\theta}_2 \mathbf{S}_2 + \dot{\theta}_1 \dot{\theta}_2 \mathbf{S}_1 \wedge \mathbf{S}_2$$

So in terms of the joint angles the equations of motion become:-

$$\tau_1 = \ddot{\theta}_1 \mathbf{S}_1^T (\mathbf{N}_1 + \mathbf{N}_2) \mathbf{S}_1 + \ddot{\theta}_2 \mathbf{S}_2^T \mathbf{N}_2 \mathbf{S}_1 + \dot{\theta}_1 \dot{\theta}_2 \mathbf{S}_1^T (\mathbf{N}_1 - \mathbf{N}_2)(\mathbf{S}_1 \wedge \mathbf{S}_2)$$
$$- \dot{\theta}_2^2 \mathbf{S}_2^T \mathbf{N}_2 (\mathbf{S}_1 \wedge \mathbf{S}_2) - (\mathcal{G}_1^T + \mathcal{G}_2^T) \mathbf{S}_1$$

$$\tau_2 = \ddot{\theta}_1 \mathbf{S}_1^T \mathbf{N}_2 \mathbf{S}_2 + \ddot{\theta}_2 \mathbf{S}_2^T \mathbf{N}_2 \mathbf{S}_2 + 2\dot{\theta}_1 \dot{\theta}_2 \mathbf{S}_2^T \mathbf{N}_2 (\mathbf{S}_1 \wedge \mathbf{S}_2) + \dot{\theta}_1^2 \mathbf{S}_1^T \mathbf{N}_2 (\mathbf{S}_1 \wedge \mathbf{S}_2)$$
$$- \mathcal{G}_2^T \mathbf{S}_2$$

Unfortunately this is still not exactly the equations of motion in terms of the joint angles, since the dependence of the $\mathbf{S}$'s, the $\mathbf{N}$'s and the $\mathcal{G}$'s has not been made explicit. If we continue this the equations are going to get very long indeed. Such long formulas do not tell us very much. However in a robot control system it may be necessary to use the formulas to calculate the torques. We will see a little later how this can be done efficiently without the need to derive the equations explicitly.

9.6   Many Links

Now we have all the mathematical tools we need to write down the equations of motion for a six joint serial manipulator. First we write down the equations of motion for each link:-

$$\tau_6 + \mathcal{R}_6 + \mathcal{G}_6 = \mathbf{N}_6\dot{\mathbf{V}}_6 + \left\{\mathbf{V}_6, \mathbf{N}_6\mathbf{V}_6\right\}$$

$$\tau_5 + \mathcal{R}_5 + \mathcal{G}_5 - \tau_6 - \mathcal{R}_6 = \mathbf{N}_5\dot{\mathbf{V}}_5 + \left\{\mathbf{V}_5, \mathbf{N}_5\mathbf{V}_5\right\}$$

$$\tau_4 + \mathcal{R}_4 + \mathcal{G}_4 - \tau_5 - \mathcal{R}_5 = \mathbf{N}_4\dot{\mathbf{V}}_4 + \left\{\mathbf{V}_4, \mathbf{N}_4\mathbf{V}_4\right\}$$

$$\tau_3 + \mathcal{R}_3 + \mathcal{G}_3 - \tau_4 - \mathcal{R}_4 = \mathbf{N}_3\dot{\mathbf{V}}_3 + \left\{\mathbf{V}_3, \mathbf{N}_3\mathbf{V}_3\right\}$$

$$\tau_2 + \mathcal{R}_2 + \mathcal{G}_2 - \tau_3 - \mathcal{R}_3 = \mathbf{N}_2\dot{\mathbf{V}}_2 + \left\{\mathbf{V}_2, \mathbf{N}_2\mathbf{V}_2\right\}$$

$$\tau_1 + \mathcal{R}_1 + \mathcal{G}_1 - \tau_2 - \mathcal{R}_2 = \mathbf{N}_1\dot{\mathbf{V}}_1 + \left\{\mathbf{V}_1, \mathbf{N}_1\mathbf{V}_1\right\}$$

These can now be rearranged by adding each equation to the ones above it:-

$$\tau_6 + \mathcal{R}_6 = \mathbf{N}_6\dot{\mathbf{V}}_6 + \left\{\mathbf{V}_6, \mathbf{N}_6\mathbf{V}_6\right\} - \mathcal{G}_6$$

$$\tau_5 + \mathcal{R}_5 = \mathbf{N}_5\dot{\mathbf{V}}_5 + \mathbf{N}_6\dot{\mathbf{V}}_6 + \left\{\mathbf{V}_5, \mathbf{N}_5\mathbf{V}_5\right\} + \left\{\mathbf{V}_6, \mathbf{N}_6\mathbf{V}_6\right\} - \mathcal{G}_6 - \mathcal{G}_5$$

$$\vdots$$

$$\tau_1 + \mathcal{R}_1 = \sum_{j=1}^{6}\left[\mathbf{N}_j\dot{\mathbf{V}}_j + \left\{\mathbf{V}_j, \mathbf{N}_j\mathbf{V}_j\right\} - \mathcal{G}_j\right]$$

Pairing each equation with the relevant joint screw gives the six joint torques:-

$$\tau_i = \sum_{j=i}^{6}\left[\dot{\mathbf{V}}_j^T\mathbf{N}_j\mathbf{S}_i + \mathbf{V}_j^T\mathbf{N}_j(\mathbf{V}_j \wedge \mathbf{S}_i) - \mathcal{G}_j^T\mathbf{S}_i\right] \qquad i = 1, 2, \ldots, 6$$

The wrench due to gravity on the jth link is a force acting at the link's centre of mass. So if our co-ordinates are aligned with the z-axis pointing up, the gravity wrenches can be written:-

$$\mathcal{G}_j = \begin{pmatrix} -Mg\mathbf{c}_i \wedge \mathbf{k} \\ -Mg\mathbf{k} \end{pmatrix}$$

Consider the following screw:-

$$\mathbf{G} = \begin{pmatrix} 0 \\ -g\mathbf{k} \end{pmatrix}$$

It is simple to check that this gravity screw has the property:-

$$\mathcal{G}_j = \mathbf{N}_j \mathbf{G}$$

Using this, we can write the equations of motion very compactly as:-

$$\tau_i = \sum_{j=i}^{6} \left[(\dot{\mathbf{V}}_j - \mathbf{G})^T \mathbf{N}_j \mathbf{S}_i + \mathbf{V}_j^T \mathbf{N}_j (\mathbf{V}_j \wedge \mathbf{S}_i) \right] \qquad i = 1, 2, \ldots, 6$$

Once again, it is more usual and more useful for control purposes to write the equations in terms of the joint angles. To do this we must use the relations:-

$$\mathbf{V}_j = \sum_{k=1}^{j} \dot{\theta}_k \mathbf{S}_k$$

and:-

$$\dot{\mathbf{V}}_j = \sum_{k=1}^{j} \left[\ddot{\theta}_k \mathbf{S}_k + \sum_{l=1}^{k-1} \dot{\theta}_k \dot{\theta}_l \mathbf{S}_l \wedge \mathbf{S}_k \right]$$

Substituting these into the equations of motion gives:-

$$\begin{aligned}
\tau_i = {} & \sum_{j=i}^{6} \left[\sum_{k=1}^{j} \ddot{\theta}_k \mathbf{S}_k^T \mathbf{N}_j \mathbf{S}_i + \sum_{k=2}^{j} \sum_{l=1}^{k-1} \dot{\theta}_k \dot{\theta}_l \mathbf{S}_i^T \mathbf{N}_j (\mathbf{S}_l \wedge \mathbf{S}_k) \right. \\
& \left. + \sum_{k=1}^{j} \sum_{l=1}^{j} \dot{\theta}_k \dot{\theta}_l \mathbf{S}_k^T \mathbf{N}_j (\mathbf{S}_l \wedge \mathbf{S}_i) - \mathbf{G}^T \mathbf{N}_j \mathbf{S}_i \right]
\end{aligned}$$

When using this, it must be remembered that the screws and inertia matrices are position dependent; we should have written $\mathbf{S}_j(\boldsymbol{\theta})$ and $\mathbf{N}_j(\boldsymbol{\theta})$, rather than just $\mathbf{S}_j$ and $\mathbf{N}_j$ to emphasize the point.

Sometimes it is useful to write the equations in the form:-

$$\tau_i = \sum_{j=1}^{6} \mathbf{A}_{ij} \ddot{\theta}_j + \sum_{j=1}^{6} \sum_{k=1}^{6} \mathbf{B}_{ijk} \dot{\theta}_j \dot{\theta}_k + \mathbf{C}_i$$

The matrix $\mathbf{A}_{ij}$ is the inertia matrix of the combined system, while $\mathbf{B}_{ijk}$ contains the information about the interaction between the links. Finally, $\mathbf{C}_i$ are the weights of the links. It is straightforward, but not very illuminating, to find these functions from the equations above.

9.7 Recursive Equations of Motion

In the previous section we derived the general equations of motion for a robot with six links connected in series. However, these results are not particularly useful for controlling a robot since although they look reasonably simple, they are difficult to compute. The problem we are faced with is how to calculate the required torques given; the joint angles, and their first and second derivatives together with the home values of the inertia matrices and joint screws. The difficulty is that the equations are written in terms of the current values of the joint screws and the inertia matrices. So we would first have to use the forward kinematics to find the current values and then substitute into the formulas. As an algorithm this is extremely inefficient and in a real system would take far too long to be of any use.

A more efficient algorithm relies on two improvements. To begin with, calculations can be done in co-ordinates at rest relative to the links, so that the link inertias will be constant in these co-ordinates. Secondly, the co-ordinate transformations that must be done can be computed recursively.

This seems to be the only place where it is more convenient to use lots of co-ordinate frames. It does mean that we must introduce rather a lot of new notation. We already know that the current values for the joint screws are given by:-

$$\mathbf{S}_i(\boldsymbol{\theta}) = \mathbf{H}_1(\theta_1)\mathbf{H}_2(\theta_2)\cdots\mathbf{H}_i(\theta_i)\mathbf{S}_i(0)$$

Here the $\mathbf{H}$ matrices are the 6×6 analogues of the $\mathbf{A}$ matrices of chapter 4. We will usually assume their dependence on joint angles and simply write them as $\mathbf{H}_i$. We will also abbreviate the home positions of the joint screws to $\mathbf{S}_i^0$. Now the velocity screws of the links can be written:-

$$\mathbf{V}_i = \mathbf{H}_1\mathbf{H}_2\cdots\mathbf{H}_i\mathbf{V}_i^0$$

where $\mathbf{V}_i^0$ is the velocity of the ith link expressed in co-ordinates at rest with respect to that link. Similarly, for the inertia matrices we have:-

$$\mathbf{N}_i = (\mathbf{H}_1^T)^{-1}(\mathbf{H}_2^T)^{-1}\cdots(\mathbf{H}_i^T)^{-1}\mathbf{N}_i^0\mathbf{H}_i^{-1}\cdots\mathbf{H}_2^{-1}\mathbf{H}_1^{-1}$$

Again, $\mathbf{N}_i^0$ represents the value of the ith link's inertia matrix in co-ordinates at rest with respect to the ith link. The point of all this is that the $\mathbf{S}_i^0$ and $\mathbf{N}_i^0$ are constants which can be predetermined.

As we saw in the previous section the torques are given by:-

$$\tau_i = \sum_{j=i}^{6}\left[(\dot{\mathbf{V}}_j - \mathbf{G})^T\mathbf{N}_j\mathbf{S}_i + \mathbf{V}_j^T\mathbf{N}_j(\mathbf{V}_j \wedge \mathbf{S}_i)\right] \qquad i = 1, 2, \ldots, 6$$

If we write:-

$$\mathcal{Q}_j = \mathbf{N}_j(\dot{\mathbf{V}}_j - \mathbf{G}) + \left\{\mathbf{V}_j, \mathbf{N}_j\mathbf{V}_j\right\}$$

then the torques become

$$\tau_i = \sum_{j=i}^{6}\mathcal{Q}_j^T\mathbf{S}_i \qquad i = 1, 2, \ldots, 6$$

As long as the wrench $\mathcal{Q}_j$ and the screw $\mathbf{S}_i$ are calculated in the same co-ordinate system, the result of the pairing $\mathcal{Q}_j^T\mathbf{S}_i$ will be independent of co-ordinates. So we will calculate $\mathcal{Q}_j^0$, the value of the wrench in co-ordinates at rest, relative to the jth link. This is just a matter of calculating:-

$$\mathcal{Q}_j^0 = \mathbf{N}_j^0(\dot{\mathbf{V}}_j^0 - \mathbf{G}_j^0) + \left\{ \mathbf{V}_j^0, \mathbf{N}_j^0 \mathbf{V}_j^0 \right\}$$

Notice that the constant gravity screw $\mathbf{G}$, must also be transformed:-

$$\mathbf{G} = \mathbf{H}_1 \mathbf{H}_2 \cdots \mathbf{H}_j \mathbf{G}_j^0$$

The velocity of the links is given by:-

$$\mathbf{V}_j = \dot{\theta}_1 \mathbf{S}_1 + \dot{\theta}_2 \mathbf{S}_2 + \cdots + \dot{\theta}_j \mathbf{S}_j$$

In terms of the co-ordinates at rest relative to the jth link, this becomes:-

$$\mathbf{V}_j^0 = \dot{\theta}_1 \mathbf{H}_j^{-1} \cdots \mathbf{H}_2^{-1} \mathbf{S}_1^0 + \dot{\theta}_2 \mathbf{H}_j^{-1} \cdots \mathbf{H}_3^{-1} \mathbf{S}_2^0 + \cdots + \dot{\theta}_j \mathbf{S}_j^0$$

It seems that we must multiply many matrices here, but in fact we can reduce the computational burden considerably by using a recursive procedure:-

$$\mathbf{V}_j^0 = \mathbf{H}_j^{-1} \mathbf{V}_{j-1}^0 + \dot{\theta}_j \mathbf{S}_j^0; \qquad \mathbf{V}_1^0 = \dot{\theta}_1 \mathbf{S}_1^0$$

This now requires just six multiplications of six component vectors by matrices to get all six velocities. The accelerations can be treated in a similar fashion. The acceleration of the jth link is given by:-

$$\dot{\mathbf{V}}_j = \ddot{\theta}_1 \mathbf{S}_1 + \ddot{\theta}_2 \mathbf{S}_2 + \cdots + \ddot{\theta}_j \mathbf{S}_j + \dot{\theta}_2 \mathbf{V}_2 \wedge \mathbf{S}_2 + \dot{\theta}_3 \mathbf{V}_3 \wedge \mathbf{S}_3 + \cdots + \dot{\theta}_j \mathbf{V}_j \wedge \mathbf{S}_j$$

Using the co-ordinates at rest relative to the jth link this becomes:-

$$\begin{aligned} \dot{\mathbf{V}}_j^0 &= \ddot{\theta}_1 \mathbf{H}_j^{-1} \cdots \mathbf{H}_2^{-1} \mathbf{S}_1^0 + \cdots + \ddot{\theta}_{j-1} \mathbf{H}_j^{-1} \mathbf{S}_{j-1}^0 + \ddot{\theta}_j \mathbf{S}_j^0 \\ &+ \dot{\theta}_2 \mathbf{H}_j^{-1} \cdots \mathbf{H}_3^{-1}(\mathbf{V}_2^0 \wedge \mathbf{S}_2^0) + \cdots + \dot{\theta}_{j-1} \mathbf{H}_j^{-1}(\mathbf{V}_{j-1}^0 \wedge \mathbf{S}_{j-1}^0) + \dot{\theta}_j(\mathbf{V}_j^0 \wedge \mathbf{S}_j^0) \end{aligned}$$

This can be calculated with the recursive scheme:-

$$\dot{\mathbf{V}}_j^0 = \mathbf{H}_j^{-1} \dot{\mathbf{V}}_{j-1}^0 + \ddot{\theta}_j \mathbf{S}_j^0 + \dot{\theta}_j(\mathbf{V}_j^0 \wedge \mathbf{S}_j^0); \qquad \dot{\mathbf{V}}_1^0 = \ddot{\theta}_1 \mathbf{S}_1^0$$

We can do even better by including the gravity screw:-

$$(\dot{\mathbf{V}}_j^0 - \mathbf{G}_j^0) = \mathbf{H}_j^{-1}(\dot{\mathbf{V}}_{j-1}^0 - \mathbf{G}_{j-1}^0) + \ddot{\theta}_j \mathbf{S}_j^0 + \dot{\theta}_j(\mathbf{V}_j^0 \wedge \mathbf{S}_j^0); \qquad \dot{\mathbf{V}}_1^0 - \mathbf{G}_1^0 = \ddot{\theta}_1 \mathbf{S}_1^0 - \mathbf{G}$$

Again only six matrix multiplications are needed. These results can now be assembled and the wrenches $\mathcal{Q}_j^0$ found. Unfortunately these cannot be simply paired with the joint screws $\mathbf{S}_i^0$, since they refer to different co-ordinate frames. Let us write:-

$$\mathcal{P}_i = \sum_{j=i}^{6} \mathcal{Q}_j$$

so that the torques will be:-

$$\tau_i = \mathcal{P}_i^T \mathbf{S}_i = (\mathcal{P}_i^0)^T \mathbf{S}_i^0$$

Using the co-ordinates at rest relative to the ith link gives:-

$$\mathcal{P}_i^0 = (\mathbf{H}_{i+1}^T)^{-1}(\mathbf{H}_{i+2}^T)^{-1} \cdots (\mathbf{H}_6^T)^{-1} \mathcal{Q}_6^0 + (\mathbf{H}_{i+1}^T)^{-1}(\mathbf{H}_{i+2}^T)^{-1} \cdots (\mathbf{H}_5^T)^{-1} \mathcal{Q}_5^0 + \cdots + \mathcal{Q}_i^0$$

These quantities can also be evaluated recursively, this time beginning with the last one:-

$$\mathcal{P}_i^0 = (\mathbf{H}_{i+1}^T)^{-1}\mathcal{P}_{i+1}^0 + \mathcal{Q}_i^0; \qquad \mathcal{P}_6^0 = \mathcal{Q}_6^0$$

Notice that although this procedure has been derived using co-ordinate frames at rest with respect to the links, we could view the process as a purely algebraic one, involving the **H** matrices; that is, without referring to different co-ordinate frames.

To summarize, the algorithm can be written in stages:-

- Calculate the six link velocities;

$$\mathbf{V}_j^0 = \mathbf{H}_j^{-1}\mathbf{V}_{j-1}^0 + \dot{\theta}_j\mathbf{S}_j^0; \qquad \mathbf{V}_1^0 = \dot{\theta}_1\mathbf{S}_1^0$$

- Calculate the six link accelerations;

$$(\dot{\mathbf{V}}_j^0 - \mathbf{G}_j^0) = \mathbf{H}_j^{-1}(\dot{\mathbf{V}}_{j-1}^0 - \mathbf{G}_{j-1}^0) + \ddot{\theta}_j\mathbf{S}_j^0 + \dot{\theta}_j(\mathbf{V}_j^0 \wedge \mathbf{S}_j^0); \qquad \dot{\mathbf{V}}_1^0 - \mathbf{G}_1^0 = \ddot{\theta}_1\mathbf{S}_1^0 - \mathbf{G}$$

- Calculate the wrench due to each link;

$$\mathcal{Q}_j^0 = \mathbf{N}_j^0(\dot{\mathbf{V}}_j^0 - \mathbf{G}_j^0) + \left\{\mathbf{V}_j^0, \mathbf{N}_j^0\mathbf{V}_j^0\right\}$$

- Calculate the total wrench acting on each link;

$$\mathcal{P}_i^0 = (\mathbf{H}_{i+1}^T)^{-1}\mathcal{P}_{i+1}^0 + \mathcal{Q}_i^0; \qquad \mathcal{P}_6^0 = \mathcal{Q}_6^0$$

- Calculate the torques by pairing with the joint screws;

$$\tau_i = (\mathcal{P}_i^0)^T\mathbf{S}_i^0$$

The calculations outlined above must be done as efficiently as possible; for example, using the 6×6 matrix **N** it is best to use the partitioned form:-

$$\mathbf{N}\mathbf{V} = \begin{pmatrix} \mathbf{I} & M\mathbf{C} \\ M\mathbf{C}^T & M\mathbf{I} \end{pmatrix}\begin{pmatrix} \omega \\ \mathbf{s} \end{pmatrix} = \begin{pmatrix} \mathbf{I}\omega + M\mathbf{c} \wedge \mathbf{s} \\ M\mathbf{s} - M\mathbf{c} \wedge \omega \end{pmatrix}$$

This needs only 24 multiplications as opposed to the 216 required for a general 6×6 matrix multiplication. Hence, using this scheme a new set of torques can be computed in something like a millisecond. If this is still too long, the process can be speeded up using several processors working in parallel.

In this final chapter we have seen how the dynamics of simple open loop robot manipulators can be written down in a fairly straightforward manner. To be useful for practical applications, however, the elements of the inertia matrices must be found. These numbers are probably best determined experimentally for each robot. We have also seen how to speed up the computation of the joint torques using a recursive algorithm.

Exercises

9.4   Consider a door hung on a helical hinge; the hinge is effectively a helical joint of pitch p, with axis pointing vertically upwards. Find the equation of motion as the door closes under its own weight.

9.5   The kinetic energy of a rigid body can be written:-

$$KE = \frac{1}{2}\mathbf{V}^T\mathbf{N}\,\mathbf{V}$$

where $\mathbf{V}$ is the velocity screw of the body and $\mathbf{N}$ its inertia matrix. Write down the kinetic energy for a serial six joint manipulator and find expressions for the partial derivatives of the kinetic energy with respect to the joint velocities; that is the differential coefficients:-

$$\frac{\partial KE}{\partial \dot{\theta}_1}, \quad \frac{\partial KE}{\partial \dot{\theta}_2}, \ldots, \frac{\partial KE}{\partial \dot{\theta}_6}$$

9.6   A simple model of friction assumes that the friction at any joint is proportional to the joint rate; that is, about any joint we should expect a frictional wrench opposing the motion with magnitude $\mu\dot{\theta}$. Write down the equations of motions for a six joint manipulator including this model of friction at each joint.

10 Solutions to Exercises

Chapter 2

2.1 (i)

$$
\begin{pmatrix} \cos\frac{\pi}{3} & -\sin\frac{\pi}{3} & 0 \\ \sin\frac{\pi}{3} & \cos\frac{\pi}{3} & 0 \\ 0 & 0 & 1 \end{pmatrix} = \begin{pmatrix} \frac{1}{2} & -\frac{\sqrt{3}}{2} & 0 \\ \frac{\sqrt{3}}{2} & \frac{1}{2} & 0 \\ 0 & 0 & 1 \end{pmatrix}
$$

2.1 (ii)

$$
\begin{pmatrix} \frac{1}{2} & -\frac{\sqrt{3}}{2} & 0 \\ \frac{\sqrt{3}}{2} & \frac{1}{2} & 0 \\ 0 & 0 & 1 \end{pmatrix} \begin{pmatrix} 1 & 0 & 1 \\ 0 & 1 & 0 \\ 0 & 0 & 1 \end{pmatrix} = \begin{pmatrix} \frac{1}{2} & -\frac{\sqrt{3}}{2} & \frac{1}{2} \\ \frac{\sqrt{3}}{2} & \frac{1}{2} & \frac{\sqrt{3}}{2} \\ 0 & 0 & 1 \end{pmatrix}
$$

2.1 (iii)

$$
\begin{pmatrix} 1 & 0 & 1 \\ 0 & 1 & 1 \\ 0 & 0 & 1 \end{pmatrix} \begin{pmatrix} \frac{1}{2} & -\frac{\sqrt{3}}{2} & 0 \\ \frac{\sqrt{3}}{2} & \frac{1}{2} & 0 \\ 0 & 0 & 1 \end{pmatrix} \begin{pmatrix} 1 & 0 & -1 \\ 0 & 1 & -1 \\ 0 & 0 & 1 \end{pmatrix} = \begin{pmatrix} \frac{1}{2} & -\frac{\sqrt{3}}{2} & \frac{1}{2}+\frac{\sqrt{3}}{2} \\ \frac{\sqrt{3}}{2} & \frac{1}{2} & \frac{1}{2}-\frac{\sqrt{3}}{2} \\ 0 & 0 & 1 \end{pmatrix}
$$

2.2   Centre of rotation satisfies $(\mathbf{R} - \mathbf{I})\mathbf{p} = -\mathbf{t}$.

2.2 (i)

$$
\begin{pmatrix} \frac{1}{\sqrt{2}} - 1 & -\frac{1}{\sqrt{2}} \\ \frac{1}{\sqrt{2}} & \frac{1}{\sqrt{2}} - 1 \end{pmatrix} \begin{pmatrix} p_x \\ p_y \end{pmatrix} = \begin{pmatrix} \frac{1}{\sqrt{2}} - 1 \\ \frac{1}{\sqrt{2}} \end{pmatrix}
$$

$$
p_x = 1, \qquad p_y = 0
$$

2.2 (ii)

$$
\begin{pmatrix} \frac{1}{\sqrt{2}} - 1 & -\frac{1}{\sqrt{2}} \\ \frac{1}{\sqrt{2}} & \frac{1}{\sqrt{2}} - 1 \end{pmatrix} \begin{pmatrix} p_x \\ p_y \end{pmatrix} = \begin{pmatrix} \frac{1}{\sqrt{2}} - 2 \\ \frac{3}{\sqrt{2}} - 1 \end{pmatrix}
$$

$$p_x = 2, \qquad p_y = 1$$

2.2 (iii)

$$\begin{pmatrix} -\frac{1}{2} & -\frac{\sqrt{3}}{2} \\ \frac{\sqrt{3}}{2} & -\frac{1}{2} \end{pmatrix} \begin{pmatrix} p_x \\ p_y \end{pmatrix} = \begin{pmatrix} \frac{-1-\sqrt{3}}{2} \\ \frac{-1+\sqrt{3}}{2} \end{pmatrix}$$

$$p_x = 1, \qquad p_y = 1$$

2.3   Let the unknown matrix be $\begin{pmatrix} a & -b & t_x \\ b & a & t_y \\ 0 & 0 & 1 \end{pmatrix}$, so that:-

$$\begin{pmatrix} a & -b & t_x \\ b & a & t_y \\ 0 & 0 & 1 \end{pmatrix} \begin{pmatrix} 0 \\ 1 \\ 1 \end{pmatrix} = \begin{pmatrix} \frac{1-\sqrt{3}}{2} \\ \frac{1-\sqrt{3}}{2} \\ 1 \end{pmatrix}$$

$$\begin{pmatrix} a & -b & t_x \\ b & a & t_y \\ 0 & 0 & 1 \end{pmatrix} \begin{pmatrix} 1 \\ 1 \\ 1 \end{pmatrix} = \begin{pmatrix} \frac{2-\sqrt{3}}{2} \\ \frac{1}{2} \\ 1 \end{pmatrix}$$

This gives four equations for the unknown matrix entries:-

$$\left. \begin{array}{rcl} -\ b\ +\ t_x & = & \frac{1-\sqrt{3}}{2} \\ a \qquad\quad +\ t_y & = & \frac{1-\sqrt{3}}{2} \\ a\ -\ b\ +\ t_x & = & \frac{2-\sqrt{3}}{2} \\ a\ +\ b\ +\ t_y & = & \frac{1}{2} \end{array} \right\}$$

Solving these equations gives:-

$$a = \frac{1}{2}, \quad b = \frac{\sqrt{3}}{2}, \quad t_x = \frac{1}{2}, \quad t_y = -\frac{\sqrt{3}}{2}$$

2.4   Rotating about the x-axis, then the y-axis is given by the following matrix multiplication:-

$$\begin{pmatrix} 0 & 0 & 1 \\ 0 & 1 & 0 \\ -1 & 0 & 0 \end{pmatrix} \begin{pmatrix} 1 & 0 & 0 \\ 0 & 0 & -1 \\ 0 & 1 & 0 \end{pmatrix} = \begin{pmatrix} 0 & 1 & 0 \\ 0 & 0 & -1 \\ -1 & 0 & 0 \end{pmatrix}$$

The axis of rotation is given by the eigenvector with eigenvalue 1:-

$$\begin{pmatrix} -1 & 1 & 0 \\ 0 & -1 & -1 \\ -1 & 0 & -1 \end{pmatrix} \begin{pmatrix} v_x \\ v_y \\ v_z \end{pmatrix} = \begin{pmatrix} 0 \\ 0 \\ 0 \end{pmatrix}$$

The solution of this singular system of equations is:

$$v_y = v_x, \qquad v_z = -v_x$$

Hence the unit vector along the axis of rotation is:-

$$\hat{\mathbf{v}} = \frac{1}{\sqrt{3}} \begin{pmatrix} 1 \\ 1 \\ -1 \end{pmatrix}$$

Performing the rotations in the other order gives:-

$$\begin{pmatrix} 1 & 0 & 0 \\ 0 & 0 & -1 \\ 0 & 1 & 0 \end{pmatrix} \begin{pmatrix} 0 & 0 & 1 \\ 0 & 1 & 0 \\ -1 & 0 & 0 \end{pmatrix} = \begin{pmatrix} 0 & 0 & 1 \\ 1 & 0 & 0 \\ 0 & 1 & 0 \end{pmatrix}$$

So we must solve:-

$$\begin{pmatrix} -1 & 0 & 1 \\ 1 & -1 & 0 \\ 0 & 1 & -1 \end{pmatrix} \begin{pmatrix} v'_x \\ v'_y \\ v'_z \end{pmatrix} = \begin{pmatrix} 0 \\ 0 \\ 0 \end{pmatrix}$$

$$v'_y = v'_x, \qquad v'_z = v_x$$

$$\hat{\mathbf{v}} = \frac{1}{\sqrt{3}} \begin{pmatrix} 1 \\ 1 \\ 1 \end{pmatrix}$$

2.5 (i)

$$\begin{pmatrix} 1 & 0 & 0 & 0 \\ 0 & 1 & 0 & 0 \\ 0 & 0 & 1 & 2 \\ 0 & 0 & 0 & 1 \end{pmatrix} \begin{pmatrix} 1 & 0 & 0 & 0 \\ 0 & \frac{1}{\sqrt{2}} & -\frac{1}{\sqrt{2}} & 0 \\ 0 & \frac{1}{\sqrt{2}} & \frac{1}{\sqrt{2}} & 0 \\ 0 & 0 & 0 & 1 \end{pmatrix} = \begin{pmatrix} 1 & 0 & 0 & 0 \\ 0 & \frac{1}{\sqrt{2}} & -\frac{1}{\sqrt{2}} & 0 \\ 0 & \frac{1}{\sqrt{2}} & \frac{1}{\sqrt{2}} & 2 \\ 0 & 0 & 0 & 1 \end{pmatrix}$$

2.5 (ii)

$$\begin{pmatrix} 1 & 0 & 0 & 0 \\ 0 & \frac{1}{\sqrt{2}} & -\frac{1}{\sqrt{2}} & 0 \\ 0 & \frac{1}{\sqrt{2}} & \frac{1}{\sqrt{2}} & 0 \\ 0 & 0 & 0 & 1 \end{pmatrix} \begin{pmatrix} 1 & 0 & 0 & 0 \\ 0 & 1 & 0 & 0 \\ 0 & 0 & 1 & 2 \\ 0 & 0 & 0 & 1 \end{pmatrix} = \begin{pmatrix} 1 & 0 & 0 & 0 \\ 0 & \frac{1}{\sqrt{2}} & -\frac{1}{\sqrt{2}} & -\sqrt{2} \\ 0 & \frac{1}{\sqrt{2}} & \frac{1}{\sqrt{2}} & \sqrt{2} \\ 0 & 0 & 0 & 1 \end{pmatrix}$$

2.5 (iii)

$$
\begin{pmatrix} 1 & 0 & 0 & 0 \\ 0 & 1 & 0 & 0 \\ 0 & 0 & 1 & -2 \\ 0 & 0 & 0 & 1 \end{pmatrix}
\begin{pmatrix} 1 & 0 & 0 & 0 \\ 0 & \frac{1}{\sqrt{2}} & -\frac{1}{\sqrt{2}} & -\sqrt{2} \\ 0 & \frac{1}{\sqrt{2}} & \frac{1}{\sqrt{2}} & \sqrt{2} \\ 0 & 0 & 0 & 1 \end{pmatrix}
=
\begin{pmatrix} 1 & 0 & 0 & 0 \\ 0 & \frac{1}{\sqrt{2}} & -\frac{1}{\sqrt{2}} & -\sqrt{2} \\ 0 & \frac{1}{\sqrt{2}} & \frac{1}{\sqrt{2}} & \sqrt{2}-2 \\ 0 & 0 & 0 & 1 \end{pmatrix}
$$

2.6   Suppose the matrix is given by:-

$$
\begin{pmatrix} u_x & v_x & w_x & t_x \\ u_y & v_y & w_y & t_y \\ u_z & v_z & w_z & t_z \\ 0 & 0 & 0 & 1 \end{pmatrix}
$$

The first point then tells us that:-

$$
\begin{pmatrix} u_x & v_x & w_x & t_x \\ u_y & v_y & w_y & t_y \\ u_z & v_z & w_z & t_z \\ 0 & 0 & 0 & 1 \end{pmatrix}
\begin{pmatrix} 0 \\ 0 \\ 0 \\ 1 \end{pmatrix}
=
\begin{pmatrix} 1 \\ -1 \\ 1 \\ 1 \end{pmatrix}
$$

so that $t_x = 1$, $t_y = -1$ and $t_z = 1$. The next point yields:-

$$
\begin{pmatrix} u_x & v_x & w_x & 1 \\ u_y & v_y & w_y & -1 \\ u_z & v_z & w_z & 1 \\ 0 & 0 & 0 & 1 \end{pmatrix}
\begin{pmatrix} 0 \\ 1 \\ 0 \\ 1 \end{pmatrix}
=
\begin{pmatrix} 0 \\ -1 \\ 1 \\ 1 \end{pmatrix}
$$

So $v_x = -1$, $v_y = 0$, $v_z = 0$. To find the w's we must invoke the relations $\mathbf{R}^T\mathbf{R} = \mathbf{I}$ and $\det(\mathbf{R}) = 1$. This can be done by setting $\mathbf{w} = \mathbf{v} \wedge \mathbf{u}$ ($\mathbf{w} = \mathbf{u} \wedge \mathbf{v}$ would give $\det(\mathbf{R}) = -1$):

$$
\begin{aligned}
w_x &= v_z u_y - v_y u_z &&= 0 \\
w_y &= v_x u_z - v_z u_x &&= 0 \\
w_z &= v_y u_x - v_x u_y &&= 1
\end{aligned}
$$

The 4×4 matrix is thus:-

$$
\begin{pmatrix} 0 & -1 & 0 & 1 \\ 1 & 0 & 0 & -1 \\ 0 & 0 & 1 & 1 \\ 0 & 0 & 0 & 1 \end{pmatrix}
$$

The rotation part of this transformation is recognized as a $\pi/2$ rotation about the z-axis. The pitch is then given by:-

$$p = \frac{2\pi}{(\pi/2)} \begin{pmatrix} 0 \\ 0 \\ 1 \end{pmatrix} \cdot \begin{pmatrix} 1 \\ -1 \\ 1 \end{pmatrix} = 4$$

To find a point $\mathbf{q}$ on the axis we must solve:-

$$\begin{pmatrix} 1 & 1 & 0 \\ -1 & 1 & 0 \\ 0 & 0 & 0 \end{pmatrix} \begin{pmatrix} q_x \\ q_y \\ q_z \end{pmatrix} = \begin{pmatrix} 1 \\ -1 \\ 1 \end{pmatrix} - \begin{pmatrix} 0 \\ 0 \\ 1 \end{pmatrix} = \begin{pmatrix} 1 \\ -1 \\ 0 \end{pmatrix}$$

$q_x = 1$, $q_y = 0$ and $q_z =$ anything. The point on the axis satisfying $\mathbf{q} \cdot \hat{\mathbf{v}} = 0$ is:-

$$\mathbf{q} = \begin{pmatrix} 1 \\ 0 \\ 0 \end{pmatrix}$$

2.7 (i)

$$\mathbf{A} = \mathbf{R} - \mathbf{R}^T$$

So that:-

$$\mathbf{A}^T = \mathbf{R}^T - \mathbf{R} = -(\mathbf{R} - \mathbf{R}^T) = -\mathbf{A}$$

2.7 (ii)

$$\mathbf{A} = \mathbf{0} \Leftrightarrow \mathbf{R} = \mathbf{R}^T$$

Now $\mathbf{R}^T\mathbf{R} = \mathbf{I}$ in general, here we have $\mathbf{R}^2 = \mathbf{I}$. If $\mathbf{R} = \mathbf{R}(\theta, \hat{\mathbf{v}})$ then $\mathbf{R}^2 = \mathbf{R}(2\theta, \hat{\mathbf{v}})$, and hence $2\theta = 2n\pi$. The even values of n give $\theta = 0$ and odd n yields $\theta = \pi$.

2.7 (iii) $\mathbf{R}\mathbf{v} = \gamma\mathbf{v}$ so $\mathbf{v} = \frac{1}{\gamma}\mathbf{R}\mathbf{v}$

Now $\mathbf{R}^T\mathbf{v} = \mathbf{R}^T(\frac{1}{\gamma}\mathbf{R})\mathbf{v} = \frac{1}{\gamma}(\mathbf{R}^T\mathbf{R})\mathbf{v} = \frac{1}{\gamma}\mathbf{v}$

2.7 (iv) The matrix $\mathbf{A}$ is a 3×3 antisymmetric matrix, so we may write it as:-

$$\mathbf{A} = \begin{pmatrix} 0 & a_z & -a_y \\ -a_z & 0 & a_x \\ a_y & -a_x & 0 \end{pmatrix}$$

For any vector $\mathbf{u} = \begin{pmatrix} u_x \\ u_y \\ u_z \end{pmatrix}$ we have:-

$$\mathbf{A}\mathbf{u} = \mathbf{a} \wedge \mathbf{u}$$

But $\mathbf{A}\mathbf{v} = \mathbf{R}\mathbf{v} - \mathbf{R}^T\mathbf{v} = \mathbf{v} - \mathbf{v} = 0$, since the eigenvalue associated with $\mathbf{v}$ is 1. This means that $\mathbf{a} \wedge \mathbf{v} = 0$ and hence $\mathbf{a} = \lambda\mathbf{v}$. To show that $\lambda = 2\sin\theta$, observe that:

$$Tr(\mathbf{A}^2) = -2(a_x^2 + a_y^2 + a_z^2)$$

where Tr denotes the trace of the matrix, that is the sum of the diagonal elements, hence:-

$$
\begin{aligned}
-2\lambda^2 &= Tr(\{\mathbf{R} - \mathbf{R}^T\}^2) \\
&= Tr(\mathbf{R}^2 - (\mathbf{R}^T)^2 - 2\mathbf{I}) \\
&= 2Tr(\mathbf{R}^2) - 6
\end{aligned}
$$

Now, if we perform a similarity transformation on $\mathbf{R}^2$, its trace will be unaffected, since for any two matrices $Tr(AB) = Tr(BA)$; that is $Tr(\mathbf{B}\mathbf{R}^2\mathbf{B}^{-1})$ $= Tr(\mathbf{B}^{-1}\mathbf{B}\mathbf{R}^2) = Tr(\mathbf{R}^2)$. This means that it does not matter which $\mathbf{R}$ we choose to do the calculation; for example a rotation about one of the co-ordinate axes would give:-

$$Tr(\mathbf{R}^2) = 1 + 2\cos 2\theta$$

So that:-

$$\lambda^2 = \frac{-1}{2}(4\cos 2\theta - 4) = 2 - 2\cos 2\theta = 4\sin^2\theta$$

Finally, we choose the positive sign for the square root, since we already have a sign ambiguity for the direction of $\hat{\mathbf{v}}$.

Chapter 3

3.1 (i)

$$
\begin{pmatrix} r_{1x} & r_{2x} & r_{3x} & t_x \\ r_{1y} & r_{2y} & r_{3y} & t_y \\ r_{1z} & r_{2z} & r_{3z} & t_z \\ 0 & 0 & 0 & 1 \end{pmatrix}
\begin{pmatrix} 0 \\ 0 \\ 0 \\ 1 \end{pmatrix}
=
\begin{pmatrix} 2 \\ 0 \\ 0 \\ 1 \end{pmatrix}
\Rightarrow
\begin{pmatrix} t_x \\ t_y \\ t_z \end{pmatrix}
=
\begin{pmatrix} 2 \\ 0 \\ 0 \end{pmatrix}
$$

$$
\begin{pmatrix} r_{1x} & r_{2x} & r_{3x} & 2 \\ r_{1y} & r_{2y} & r_{3y} & 0 \\ r_{1z} & r_{2z} & r_{3z} & 0 \\ 0 & 0 & 0 & 1 \end{pmatrix}
\begin{pmatrix} 1 \\ 0 \\ 0 \\ 1 \end{pmatrix}
=
\begin{pmatrix} 3 \\ 0 \\ 0 \\ 1 \end{pmatrix}
\Rightarrow
\begin{pmatrix} r_{1x} \\ r_{1y} \\ r_{1z} \end{pmatrix}
=
\begin{pmatrix} 1 \\ 0 \\ 0 \end{pmatrix}
$$

$$
\begin{pmatrix} 1 & r_{2x} & r_{3x} & 2 \\ 0 & r_{2y} & r_{3y} & 0 \\ 0 & r_{2z} & r_{3z} & 0 \\ 0 & 0 & 0 & 1 \end{pmatrix}
\begin{pmatrix} 0 \\ 1 \\ 0 \\ 1 \end{pmatrix}
=
\begin{pmatrix} 2 \\ 0 \\ 1 \\ 1 \end{pmatrix}
\Rightarrow
\begin{pmatrix} r_{2x} \\ r_{2y} \\ r_{2z} \end{pmatrix}
=
\begin{pmatrix} 0 \\ 0 \\ 1 \end{pmatrix}
$$

The third column is found by imposing the relations $\mathbf{R}^T\mathbf{R} = \mathbf{I}$ and $\det \mathbf{R} = 1$. This can be done by setting $\mathbf{r}_3 = \mathbf{r}_1 \wedge \mathbf{r}_2$

$$\begin{pmatrix} r_{3x} \\ r_{3y} \\ r_{3z} \end{pmatrix} = \begin{pmatrix} 0 \\ -1 \\ 0 \end{pmatrix}$$

Finally, by inspection, the motion is a screw motion about the x-axis, the angle of rotation is $\pi/2$ and the pitch is 8.

3.1 (ii)

$$\begin{pmatrix} t_x \\ t_y \\ t_z \end{pmatrix} = \begin{pmatrix} 0 \\ 0 \\ 0 \end{pmatrix},$$

$$\begin{pmatrix} r_{1x} \\ r_{1y} \\ r_{1z} \end{pmatrix} = \begin{pmatrix} 0 \\ 1 \\ 0 \end{pmatrix}, \quad \begin{pmatrix} r_{2x} \\ r_{2y} \\ r_{2z} \end{pmatrix} = \begin{pmatrix} -1 \\ 0 \\ 0 \end{pmatrix}, \quad \begin{pmatrix} r_{3x} \\ r_{3y} \\ r_{3z} \end{pmatrix} = \begin{pmatrix} 0 \\ 0 \\ 1 \end{pmatrix}$$

This is a $\pi/2$ rotation about the z-axis.

3.1 (iii)

$$\begin{pmatrix} t_x \\ t_y \\ t_z \end{pmatrix} = \begin{pmatrix} 0 \\ 0 \\ 1 \end{pmatrix},$$

$$\begin{pmatrix} r_{1x} \\ r_{1y} \\ r_{1z} \end{pmatrix} = \begin{pmatrix} 0 \\ 1 \\ 0 \end{pmatrix}, \quad \begin{pmatrix} r_{2x} \\ r_{2y} \\ r_{2z} \end{pmatrix} = \begin{pmatrix} -1 \\ 0 \\ 0 \end{pmatrix}, \quad \begin{pmatrix} r_{3x} \\ r_{3y} \\ r_{3z} \end{pmatrix} = \begin{pmatrix} 0 \\ 0 \\ 1 \end{pmatrix}$$

The motion is a screw motion about the z-axis, the angle of rotation is $\pi/2$ and the pitch is 4.

3.2 (i)

$$\begin{pmatrix} \cos\theta & -\sin\theta & 0 & 0 \\ \sin\theta & \cos\theta & 0 & 0 \\ 0 & 0 & 1 & 0 \\ 0 & 0 & 0 & 1 \end{pmatrix}$$

3.2 (ii)

$$\begin{pmatrix} 1 & 0 & 0 & 0 \\ 0 & 1 & 0 & 0 \\ 0 & 0 & 1 & d \\ 0 & 0 & 0 & 1 \end{pmatrix}$$

3.2 (iii)

$$\begin{pmatrix} 1 & 0 & 0 & 1 \\ 0 & 1 & 0 & 0 \\ 0 & 0 & 1 & 0 \\ 0 & 0 & 0 & 1 \end{pmatrix} \begin{pmatrix} \cos\theta & -\sin\theta & 0 & 0 \\ \sin\theta & \cos\theta & 0 & 0 \\ 0 & 0 & 1 & 0 \\ 0 & 0 & 0 & 1 \end{pmatrix} \begin{pmatrix} 1 & 0 & 0 & -1 \\ 0 & 1 & 0 & 0 \\ 0 & 0 & 1 & 0 \\ 0 & 0 & 0 & 1 \end{pmatrix} =$$

$$\begin{pmatrix} \cos\theta & -\sin\theta & 0 & 1-\cos\theta \\ \sin\theta & \cos\theta & 0 & -\sin\theta \\ 0 & 0 & 1 & 0 \\ 0 & 0 & 0 & 1 \end{pmatrix}$$

3.2 (iv)

$$\begin{pmatrix} 1 & 0 & 0 & 1 \\ 0 & 1 & 0 & 0 \\ 0 & 0 & 1 & 0 \\ 0 & 0 & 0 & 1 \end{pmatrix} \begin{pmatrix} \cos\theta & -\sin\theta & 0 & 0 \\ \sin\theta & \cos\theta & 0 & 0 \\ 0 & 0 & 1 & \theta/\pi \\ 0 & 0 & 0 & 1 \end{pmatrix} \begin{pmatrix} 1 & 0 & 0 & -1 \\ 0 & 1 & 0 & 0 \\ 0 & 0 & 1 & 0 \\ 0 & 0 & 0 & 1 \end{pmatrix} =$$

$$\begin{pmatrix} \cos\theta & -\sin\theta & 0 & 1-\cos\theta \\ \sin\theta & \cos\theta & 0 & -\sin\theta \\ 0 & 0 & 1 & \theta/\pi \\ 0 & 0 & 0 & 1 \end{pmatrix}$$

3.3

$$\mathbf{A}^{-1}\mathbf{A} = \left(\begin{array}{c|c} \mathbf{R}^T & -\mathbf{R}^T\mathbf{v} \\ \hline 0 & 1 \end{array}\right) \left(\begin{array}{c|c} \mathbf{R} & \mathbf{v} \\ \hline 0 & 1 \end{array}\right)$$

$$= \left(\begin{array}{c|c} \mathbf{R}^T\mathbf{R} & \mathbf{R}^T\mathbf{v} - \mathbf{R}^T\mathbf{v} \\ \hline 0 & 1 \end{array}\right)$$

$$= \left(\begin{array}{c|c} \mathbf{I} & \mathbf{0} \\ \hline 0 & 1 \end{array}\right)$$

and

$$\mathbf{A}^{-1}\mathbf{A} = \left(\begin{array}{c|c} \mathbf{R} & \mathbf{v} \\ \hline 0 & 1 \end{array}\right) \left(\begin{array}{c|c} \mathbf{R}^T & -\mathbf{R}^T\mathbf{v} \\ \hline 0 & 1 \end{array}\right)$$

$$= \left(\begin{array}{c|c} \mathbf{R}\mathbf{R}^T & -\mathbf{R}\mathbf{R}^T\mathbf{v} + \mathbf{v} \\ \hline 0 & 1 \end{array}\right)$$

$$= \left(\begin{array}{c|c} \mathbf{I} & \mathbf{0} \\ \hline 0 & 1 \end{array}\right)$$

3.4   As we saw in section 2.6, $\mathbf{t} = (\mathbf{I} - \mathbf{R})\mathbf{u}$, so that:-

$$
\begin{aligned}
(\mathbf{I} + \mathbf{R}^T)\mathbf{t} &= (\mathbf{I} + \mathbf{R}^T)(\mathbf{I} - \mathbf{R})\mathbf{u} \\
&= (\mathbf{R}^T - \mathbf{R})\mathbf{u} \\
&= 2\sin\theta\,\mathbf{u} \wedge \hat{\mathbf{v}}
\end{aligned}
$$

by the results of exercise 2.7. Now since $\hat{\mathbf{v}}_a \wedge \hat{\mathbf{v}}_b = \hat{\mathbf{v}}$, we have:-

$$
\begin{aligned}
\hat{\mathbf{v}} \wedge \hat{\mathbf{v}}_a &= (\hat{\mathbf{v}}_a \wedge \hat{\mathbf{v}}_b) \wedge \hat{\mathbf{v}}_a = \hat{\mathbf{v}}_b(\hat{\mathbf{v}}_a \cdot \hat{\mathbf{v}}_a) - \hat{\mathbf{v}}_a(\hat{\mathbf{v}}_a \cdot \hat{\mathbf{v}}_b) = \hat{\mathbf{v}}_b \\
\hat{\mathbf{v}} \wedge \hat{\mathbf{v}}_b &= (\hat{\mathbf{v}}_a \wedge \hat{\mathbf{v}}_b) \wedge \hat{\mathbf{v}}_b = \hat{\mathbf{v}}_b(\hat{\mathbf{v}}_a \cdot \hat{\mathbf{v}}_b) - \hat{\mathbf{v}}_a(\hat{\mathbf{v}}_b \cdot \hat{\mathbf{v}}_b) = -\hat{\mathbf{v}}_a
\end{aligned}
$$

using the rule for a vector triple product. Therefore:-

$$
\hat{\mathbf{v}}_b^T(\mathbf{I} + \mathbf{R}^T)\mathbf{t} = 2\sin\theta\,\hat{\mathbf{v}}_b \cdot \mathbf{u} \wedge \hat{\mathbf{v}}
$$

Cycling the scalar triple product gives $\mathbf{u} \cdot (\hat{\mathbf{v}} \wedge \hat{\mathbf{v}}_b) = \mathbf{u} \cdot \hat{\mathbf{v}}_a$. Rearranging gives the result:-

$$
\mathbf{u} \cdot \hat{\mathbf{v}}_a = \frac{1}{2\sin\theta}\hat{\mathbf{v}}_b^T(\mathbf{I} + \mathbf{R}^T)\mathbf{t}
$$

Similarly dotting with $\hat{\mathbf{v}}_a$ gives the second result:-

$$
\mathbf{u} \cdot \hat{\mathbf{v}}_b = \frac{-1}{2\sin\theta}\hat{\mathbf{v}}_a^T(\mathbf{I} + \mathbf{R}^T)\mathbf{t}
$$

Chapter 4

4.1 (i)

$$
\begin{aligned}
x &= 2\cos(\tfrac{\pi}{6}) + 2\cos(\tfrac{\pi}{3}) + \cos(\tfrac{\pi}{2}) \quad & 1 + \sqrt{3} \\
y &= 2\sin(\tfrac{\pi}{6}) + 2\sin(\tfrac{\pi}{3}) + \sin(\tfrac{\pi}{2}) \quad & 2 + \sqrt{3} \\
\Phi &= \tfrac{\pi}{6} + \tfrac{\pi}{6} + \tfrac{\pi}{6} \quad & \tfrac{\pi}{2}
\end{aligned}
$$

4.1 (ii)

$$
\begin{aligned}
x &= 2\cos(\tfrac{\pi}{2}) + 2\cos(\tfrac{11\pi}{6}) + \cos(\tfrac{13\pi}{6}) \quad & \tfrac{3\sqrt{3}}{2} \\
y &= 2\sin(\tfrac{\pi}{2}) + 2\sin(\tfrac{11\pi}{6}) + \sin(\tfrac{13\pi}{6}) \quad & \tfrac{3}{2} \\
\Phi &= \tfrac{\pi}{2} + \tfrac{4\pi}{3} + \tfrac{\pi}{3} \quad & \tfrac{\pi}{6}
\end{aligned}
$$

4.1 (iii)

$$
\begin{aligned}
x &= 2\cos(\tfrac{-\pi}{6}) + 2\cos(\tfrac{\pi}{2}) + \cos(\tfrac{\pi}{6}) \quad & \tfrac{3\sqrt{3}}{2} \\
y &= 2\sin(\tfrac{-\pi}{6}) + 2\sin(\tfrac{\pi}{2}) + \sin(\tfrac{\pi}{6}) \quad & \tfrac{3}{2} \\
\Phi &= \tfrac{-\pi}{6} + \tfrac{2\pi}{3} + \tfrac{\pi}{3} \quad & \tfrac{\pi}{6}
\end{aligned}
$$

Notice that (ii) and (iii) give the same answer, hence they are the two different postures for this position.

4.2   Suppose the home position has joint 1 aligned along the y-axis, joint 2 lying in the xy-plane and joint 3 along the x-axis. For rotations about joint 1 we have the matrix:-

$$\mathbf{R}(\theta_1, \mathbf{j}) = \begin{pmatrix} \cos\theta_1 & 0 & \sin\theta_1 \\ 0 & 1 & 0 \\ -\sin\theta_1 & 0 & \cos\theta_1 \end{pmatrix}$$

Also for the third joint we have:-

$$\mathbf{R}(\theta_3, \mathbf{i}) = \begin{pmatrix} 1 & 0 & 0 \\ 0 & \cos\theta_3 & -\sin\theta_3 \\ 0 & \sin\theta_3 & \cos\theta_3 \end{pmatrix}$$

To find the matrix representing rotations about joint 2 we use a conjugation:-

$$
\begin{aligned}
\mathbf{R}(\theta_2, \hat{\mathbf{v}}) &= \mathbf{R}(\tfrac{\pi}{4}, \mathbf{k})\mathbf{R}(\theta_2, \mathbf{i})\mathbf{R}(-\tfrac{\pi}{4}, \mathbf{k}) \\
&= \begin{pmatrix} \frac{1}{\sqrt{2}} & -\frac{1}{\sqrt{2}} & 0 \\ \frac{1}{\sqrt{2}} & \frac{1}{\sqrt{2}} & 0 \\ 0 & 0 & 1 \end{pmatrix} \begin{pmatrix} 1 & 0 & 0 \\ 0 & \cos\theta_2 & -\sin\theta_2 \\ 0 & \sin\theta_2 & \cos\theta_2 \end{pmatrix} \begin{pmatrix} \frac{1}{\sqrt{2}} & \frac{1}{\sqrt{2}} & 0 \\ \frac{1}{\sqrt{2}} & -\frac{1}{\sqrt{2}} & 0 \\ 0 & 0 & 1 \end{pmatrix} \\
&= \begin{pmatrix} \frac{1}{2}(1+\cos\theta_2) & \frac{1}{2}(1-\cos\theta_2) & \frac{1}{\sqrt{2}}\sin\theta_2 \\ \frac{1}{2}(1-\cos\theta_2) & \frac{1}{2}(1+\cos\theta_2) & -\frac{1}{\sqrt{2}}\sin\theta_2 \\ -\frac{1}{\sqrt{2}}\sin\theta_2 & \frac{1}{\sqrt{2}}\sin\theta_2 & \cos\theta_2 \end{pmatrix}
\end{aligned}
$$

So the kinematic matrix is:-

$$\mathbf{K}(\theta_1, \theta_2, \theta_3) = \mathbf{R}(\theta_1, \mathbf{j})\mathbf{R}(\theta_2, \hat{\mathbf{v}})\mathbf{R}(\theta_3, \mathbf{i})$$

The elements of $\mathbf{K}$ are thus:-

$$
\begin{aligned}
K_{11} &= \tfrac{1}{2}\cos\theta_1(1+\cos\theta_2) - \tfrac{1}{\sqrt{2}}\sin\theta_1\sin\theta_2 \\
K_{21} &= \tfrac{1}{2}(1-\cos\theta_2) \\
K_{31} &= -\tfrac{1}{2}\sin\theta_1(1+\cos\theta_2) - \tfrac{1}{\sqrt{2}}\cos\theta_1\sin\theta_2 \\
K_{12} &= \tfrac{1}{2}\cos\theta_1(1-\cos\theta_2)\cos\theta_3 + \tfrac{1}{\sqrt{2}}\cos\theta_1\sin\theta_2\sin\theta_3 \\
&\quad + \tfrac{1}{\sqrt{2}}\sin\theta_1\sin\theta_2\cos\theta_3 + \sin\theta_1\cos\theta_2\sin\theta_3 \\
K_{22} &= \tfrac{1}{2}(1+\cos\theta_2)\cos\theta_3 - \tfrac{1}{\sqrt{2}}\sin\theta_2\sin\theta_3 \\
K_{32} &= -\tfrac{1}{2}\sin\theta_1(1-\cos\theta_2)\cos\theta_3 - \tfrac{1}{\sqrt{2}}\sin\theta_1\sin\theta_2\sin\theta_3 \\
&\quad + \tfrac{1}{\sqrt{2}}\cos\theta_1\sin\theta_2\cos\theta_3 + \cos\theta_1\cos\theta_2\sin\theta_3 \\
K_{13} &= -\tfrac{1}{2}\cos\theta_1(1-\cos\theta_2)\sin\theta_3 + \tfrac{1}{\sqrt{2}}\cos\theta_1\sin\theta_2\cos\theta_3 \\
&\quad - \tfrac{1}{\sqrt{2}}\sin\theta_1\sin\theta_2\sin\theta_3 + \sin\theta_1\cos\theta_2\cos\theta_3
\end{aligned}
$$

$$K_{23} = -\tfrac{1}{2}(1 + \cos\theta_2)\sin\theta_3 - \tfrac{1}{\sqrt{2}}\sin\theta_2\cos\theta_3$$

$$K_{33} = \tfrac{1}{2}\sin\theta_1(1 - \cos\theta_2)\sin\theta_3 - \tfrac{1}{\sqrt{2}}\sin\theta_1\sin\theta_2\cos\theta_3$$

$$- \tfrac{1}{\sqrt{2}}\cos\theta_1\sin\theta_2\sin\theta_3 + \cos\theta_1\cos\theta_2\cos\theta_3$$

4.3 Assume the home position for the Scara is when the arm is stretched out along the
x-axis with all the joints parallel to the z-axis. If we pick the origin to be on the
first joint, rotations about this joint are given by:-

$$\mathbf{A}_1(\theta_1) = \begin{pmatrix} \cos\theta_1 & -\sin\theta_1 & 0 & 0 \\ \sin\theta_1 & \cos\theta_1 & 0 & 0 \\ 0 & 0 & 1 & 0 \\ 0 & 0 & 0 & 1 \end{pmatrix}$$

For the second joint the axis is shifted in the x direction a distance l_1:-

$$\mathbf{A}_2(\theta_2) = \begin{pmatrix} \cos\theta_2 & -\sin\theta_2 & 0 & l_1(1 - \cos\theta_2) \\ \sin\theta_2 & \cos\theta_2 & 0 & -l_1\sin\theta_2 \\ 0 & 0 & 1 & 0 \\ 0 & 0 & 0 & 1 \end{pmatrix}$$

Similarly for the third joint we have:-

$$\mathbf{A}_3(\theta_3) = \begin{pmatrix} \cos\theta_3 & -\sin\theta_3 & 0 & (l_1 + l_2)(1 - \cos\theta_3) \\ \sin\theta_3 & \cos\theta_3 & 0 & -(l_1 + l_2)\sin\theta_3 \\ 0 & 0 & 1 & 0 \\ 0 & 0 & 0 & 1 \end{pmatrix}$$

The final joint is a prismatic joint along the z-axis:-

$$\mathbf{A}_4(d_4) = \begin{pmatrix} 1 & 0 & 0 & 0 \\ 0 & 1 & 0 & 0 \\ 0 & 0 & 1 & d_4 \\ 0 & 0 & 0 & 1 \end{pmatrix}$$

For the Stanford manipulator the 'A' matrices are as follows. The first joint is
aligned along the z-axis:-

$$\mathbf{A}_1(\theta_1) = \begin{pmatrix} \cos\theta_1 & -\sin\theta_1 & 0 & 0 \\ \sin\theta_1 & \cos\theta_1 & 0 & 0 \\ 0 & 0 & 1 & 0 \\ 0 & 0 & 0 & 1 \end{pmatrix}$$

The second joint is aligned along the x-axis:-

$$\mathbf{A}_2(\theta_2) = \begin{pmatrix} 1 & 0 & 0 & 0 \\ 0 & \cos\theta_2 & -\sin\theta_2 & 0 \\ 0 & \sin\theta_2 & \cos\theta_2 & 0 \\ 0 & 0 & 0 & 1 \end{pmatrix}$$

The third joint is prismatic and aligned along the z-axis again:-

$$\mathbf{A}_3(d_3) = \begin{pmatrix} 1 & 0 & 0 & 0 \\ 0 & 1 & 0 & 0 \\ 0 & 0 & 1 & d_3 \\ 0 & 0 & 0 & 1 \end{pmatrix}$$

The fourth joint is a rotation about a line parallel to the z-axis, shifted in the x direction:-

$$\mathbf{A}_4(\theta_4) = \begin{pmatrix} \cos\theta_4 & -\sin\theta_4 & 0 & l_1(1-\cos\theta_4) \\ \sin\theta_4 & \cos\theta_4 & 0 & -l_1\sin\theta_4 \\ 0 & 0 & 1 & 0 \\ 0 & 0 & 0 & 1 \end{pmatrix}$$

In the home position, when $d_3 = 0$ the second and fifth joints line up:-

$$\mathbf{A}_5(\theta_5) = \begin{pmatrix} 1 & 0 & 0 & 0 \\ 0 & \cos\theta_5 & -\sin\theta_5 & 0 \\ 0 & \sin\theta_5 & \cos\theta_5 & 0 \\ 0 & 0 & 0 & 1 \end{pmatrix}$$

Again, in the home position, the sixth and fourth joints line up:-

$$\mathbf{A}_6(\theta_6) = \begin{pmatrix} \cos\theta_6 & -\sin\theta_6 & 0 & l_1(1-\cos\theta_6) \\ \sin\theta_6 & \cos\theta_6 & 0 & -l_1\sin\theta_6 \\ 0 & 0 & 1 & 0 \\ 0 & 0 & 0 & 1 \end{pmatrix}$$

4.4   Notice that the point $\mathbf{p}$ is the wrist centre of the robot. That is, it lies on the last three joint axes, and hence will not be affected by rotations about these. The final position is thus the same as the initial position.

4.5   If $\mathbf{r}'$ is the home position of the point then the forward kinematics gives us:-

$$\mathbf{A}_1\mathbf{A}_2\mathbf{A}_3\mathbf{A}_4\mathbf{A}_5\mathbf{A}_6\mathbf{r}' = \mathbf{r}$$

Hence the home position is given by:-

$$\mathbf{r}' = \mathbf{A}_6^{-1}\mathbf{A}_5^{-1}\mathbf{A}_4^{-1}\mathbf{A}_3^{-1}\mathbf{A}_2^{-1}\mathbf{A}_1^{-1}\mathbf{r}$$

The inverse matrices of the 'A' matrices are easily calculated since:-

$$\mathbf{A}^{-1}(\theta) = \mathbf{A}(-\theta)$$

So that:-

$$\mathbf{A}_1^{-1}(\frac{\pi}{2})\mathbf{r} = \begin{pmatrix} 0 & 1 & 0 & 0 \\ -1 & 0 & 0 & 0 \\ 0 & 0 & 1 & 0 \\ 0 & 0 & 0 & 1 \end{pmatrix} \begin{pmatrix} L_2 + D_4 \\ D_3 + D_4 \\ 0 \\ 1 \end{pmatrix} = \begin{pmatrix} D_3 + D_4 \\ -L_2 - D_4 \\ 0 \\ 1 \end{pmatrix}$$

$$\mathbf{A}_2^{-1}(\frac{\pi}{2})\mathbf{A}_1^{-1}(\frac{\pi}{2})\mathbf{r} = \begin{pmatrix} 1 & 0 & 0 & 0 \\ 0 & 0 & 1 & 0 \\ 0 & -1 & 0 & 0 \\ 0 & 0 & 0 & 1 \end{pmatrix} \begin{pmatrix} D_3 + D_4 \\ -L_2 - D_4 \\ 0 \\ 1 \end{pmatrix} = \begin{pmatrix} D_3 + D_4 \\ 0 \\ L_2 + D_4 \\ 1 \end{pmatrix}$$

$$\mathbf{A}_3^{-1}(0) = \mathbf{I}$$

$$\mathbf{A}_4^{-1}(-\frac{\pi}{2})\mathbf{A}_3^{-1}(0)\mathbf{A}_2^{-1}(\frac{\pi}{2})\mathbf{A}_1^{-1}(\frac{\pi}{2})\mathbf{r}$$

$$= \begin{pmatrix} 0 & -1 & 0 & D_3 \\ 1 & 0 & 0 & -D_3 \\ 0 & 0 & 1 & 0 \\ 0 & 0 & 0 & 1 \end{pmatrix} \begin{pmatrix} D_3 + D_4 \\ 0 \\ L_2 + D_4 \\ 1 \end{pmatrix}$$

$$= \begin{pmatrix} D_3 \\ D_4 \\ L_2 + D_4 \\ 1 \end{pmatrix}$$

$$\mathbf{A}_5^{-1}(\frac{\pi}{2}) \cdots \mathbf{A}_1^{-1}(\frac{\pi}{2})\mathbf{r} = \begin{pmatrix} 1 & 0 & 0 & 0 \\ 0 & 0 & 1 & -L_2 - D_4 \\ 0 & -1 & 0 & L_2 + D_4 \\ 0 & 0 & 0 & 1 \end{pmatrix} \begin{pmatrix} D_3 \\ D_4 \\ L_2 + D_4 \\ 1 \end{pmatrix}$$

$$= \begin{pmatrix} D_3 \\ -L_2 \\ L_2 \\ 1 \end{pmatrix}$$

$$\mathbf{A}_6^{-1}(\tfrac{\pi}{2})\cdots\mathbf{A}_1^{-1}(\tfrac{\pi}{2})\mathbf{r} = \begin{pmatrix} 0 & 1 & 0 & D_3 \\ -1 & 0 & 0 & D_3 \\ 0 & 0 & 1 & 0 \\ 0 & 0 & 0 & 1 \end{pmatrix} \begin{pmatrix} D_3 \\ -L_2 \\ L_2 \\ 1 \end{pmatrix} = \begin{pmatrix} D_3 - L_2 \\ 0 \\ L_2 \\ 1 \end{pmatrix}$$

Hence the original position of the point was:-

$$\mathbf{r}' = \begin{pmatrix} D_3 - L_2 \\ 0 \\ L_2 \end{pmatrix}$$

Chapter 5

5.1 (i) Using the inverse kinematic relations we have:-

$$\cos\theta_2 = \frac{\sqrt{3}}{2}$$

$$\sin\theta_2 = \pm\frac{1}{2}$$

Hence $\theta_2 = \pm\dfrac{\pi}{6}$ radians

$$\cos\theta_1 = \frac{\sqrt{3}}{2} \text{ or } \frac{-2 + 6\sqrt{3}}{13}$$

$$\sin\theta_1 = \frac{1}{2} \text{ or } \frac{3 + 4\sqrt{3}}{13}$$

Hence $\theta_1 = \dfrac{\pi}{6}$ or 0.869 radians

5.1 (ii) This point is unreachable, since $x^2 + y^2 > (l_1 + l_2)^2$.

5.1 (iii)

$$\cos\theta_2 = \frac{1}{\sqrt{2}}$$

$$\sin\theta_2 = \pm\frac{1}{\sqrt{2}}$$

Hence $\theta_2 = \pm\dfrac{\pi}{4}$ radians

$$\cos\theta_1 = \frac{1}{\sqrt{2}} \text{ or } \frac{3}{4 + 5\sqrt{2}}$$

$$\sin\theta_1 = \frac{1}{\sqrt{2}} \text{ or } \frac{5 + 4\sqrt{2}}{4 + 5\sqrt{2}}$$

Hence $\theta_1 = \dfrac{\pi}{4}$ or 1.296 radians

5.2 (i) The inverse kinematics gives:-

$$\cos\theta_2 = \frac{\sqrt{3}}{2}, \qquad \sin\theta_2 = \pm\frac{1}{2}; \qquad \theta_2 = \pm\frac{\pi}{6} \text{ radians}$$
$$\cos\theta_1 = \pm 1, \qquad \sin\theta_1 = 0; \qquad \theta_1 = 0 \text{ or } \pi \text{ radians}$$
$$\cos\theta_3 = \pm\frac{\sqrt{3}}{2}, \qquad \sin\theta_3 = \mp\frac{1}{2}; \qquad \theta_3 = \frac{-\pi}{6} \text{ or } \frac{5\pi}{6} \text{ radians}$$

5.2 (ii)

$$\cos\theta_2 = \frac{1}{2}, \qquad \sin\theta_2 = \pm\frac{\sqrt{3}}{2}; \qquad \theta_2 = \pm\frac{\pi}{3} \text{ radians}$$
$$\cos\theta_1 = \pm\frac{\sqrt{3}}{2}, \qquad \sin\theta_1 = \pm\frac{1}{2}; \qquad \theta_1 = \frac{\pi}{6} \text{ or } \frac{7\pi}{6} \text{ radians}$$
$$\cos\theta_3 = \pm 1, \qquad \sin\theta_3 = 0; \qquad \theta_3 = 0 \text{ or } \pi \text{ radians}$$

5.2 (iii)

$$\cos\theta_2 = \frac{1}{2}, \qquad \sin\theta_2 = \pm\frac{\sqrt{3}}{2}; \qquad \theta_2 = \pm\frac{\pi}{3} \text{ radians}$$
$$\cos\theta_1 = \pm\frac{\sqrt{3}}{2}, \qquad \sin\theta_1 = \pm\frac{1}{2}; \qquad \theta_1 = \frac{\pi}{6} \text{ or } \frac{7\pi}{6} \text{ radians}$$
$$\cos\theta_3 = \pm\frac{1}{2}, \qquad \sin\theta_3 = 0; \qquad \text{results are inconsistent,}$$
$$\text{second point cannot be reached}$$

5.3   The forward kinematics gives:-

$$\begin{aligned}
x &= l_1\cos\theta_1 + l_2\cos(\theta_1+\theta_2) + l_3\cos(\theta_1+\theta_2+\theta_3) \\
y &= l_1\sin\theta_1 + l_2\sin(\theta_1+\theta_2) + l_3\sin(\theta_1+\theta_2+\theta_3) \\
\Phi &= \theta_1 + \theta_2 + \theta_3
\end{aligned}$$

By writing:-

$$\begin{aligned}
X = (x - l_3\cos\Phi) &= l_1\cos\theta_1 + l_2\cos(\theta_1+\theta_2) \\
Y = (y - l_3\sin\Phi) &= l_1\sin\theta_1 + l_2\sin(\theta_1+\theta_2)
\end{aligned}$$

we may use the results for the two joint planar manipulator given in section 5.1:-

$$\begin{aligned}
\cos\theta_2 &= \frac{1}{2l_1l_2}\{(X^2+Y^2) - (l_1^2+l_2^2)\} = \lambda \\
\sin\theta_2 &= \pm\sqrt{1-\lambda^2} \\
\cos\theta_1 &= \frac{1}{(X^2+Y^2)}\{X(l_1+l_2\lambda) \pm Yl_2\sqrt{1-\lambda^2}\} \\
\sin\theta_1 &= \frac{1}{(X^2+Y^2)}\{\mp Xl_2\sqrt{1-\lambda^2} + Y(l_1+l_2\lambda)\} \\
\theta_3 &= \Phi - \theta_2 - \theta_1
\end{aligned}$$

If we wish to place the end point at $x = 0.5$, $y = 3.0$ and $\phi = \frac{2\pi}{3}$, with $l_1 = 2, l_2 = 1$ and $l_3 = 1$, then:-

$$X = x - 1\cos\frac{2\pi}{3} = 1.0000$$

$$Y = y - 1\sin\frac{2\pi}{3} = 2.1340$$

So the computations yield:-

$$\cos\theta_2 = 0.1385$$
$$\sin\theta_2 = \pm0.9904$$
$$\cos\theta_1 = 0.7656 \text{ or } 0.0045$$
$$\sin\theta_1 = 0.6433 \text{ or } 1.0000$$

Thus; to four decimal places, the joint angles must be:-

$$\begin{pmatrix} \theta_1 \\ \theta_2 \\ \theta_3 \end{pmatrix} = \begin{pmatrix} 0.6989 \\ 1.4319 \\ -0.0364 \end{pmatrix} \text{ or } \begin{pmatrix} 1.5663 \\ -1.4319 \\ 1.9600 \end{pmatrix} \text{ radians}$$

5.4   Let us call the two points a and c, with the home positions:-

$$a = \begin{pmatrix} 0 \\ 0 \\ 1 \end{pmatrix}, \qquad c = \begin{pmatrix} 0 \\ 1 \\ 0 \end{pmatrix}$$

The forward kinematics gives the general positions of these points:-

$$\begin{pmatrix} x_a \\ y_a \\ z_a \end{pmatrix} = \begin{pmatrix} \cos\theta_1\sin\theta_2 \\ \sin\theta_1\sin\theta_2 \\ \cos\theta_2 \end{pmatrix},$$

$$\begin{pmatrix} x_c \\ y_c \\ z_c \end{pmatrix} = \begin{pmatrix} -\cos\theta_1\cos\theta_2\sin\theta_3 - \sin\theta_1\cos\theta_3 \\ -\sin\theta_1\cos\theta_2\sin\theta_3 + \cos\theta_1\cos\theta_3 \\ \sin\theta_2\sin\theta_3 \end{pmatrix}$$

Hence the inverse kinematics are given by:-

$$\cos\theta_2 = z_a$$
$$\sin\theta_2 = \pm\sqrt{1 - z_a^2}$$
$$\cos\theta_1 = x_a/\sin\theta_2$$
$$\sin\theta_1 = y_a/\sin\theta_2$$
$$\cos\theta_3 = y_c\cos\theta_1 - x_c\sin\theta_1$$
$$\sin\theta_3 = z_c/\sin\theta_2$$

5.5 (i) The calculations give:-

$\cos\theta_1$	$-7/17\sqrt{2}$	$-7/17\sqrt{2}$	$1/\sqrt{2}$	$1/\sqrt{2}$
$\sin\theta_1$	$-23/17\sqrt{2}$	$-23/17\sqrt{2}$	$1/\sqrt{2}$	$1/\sqrt{2}$
$\cos\theta_2$	-1	0	0	-1
$\sin\theta_2$	0	-1	1	0
$\cos\theta_3$	0	0	0	0
$\sin\theta_3$	1	-1	1	-1

The four postures are thus given by:-

$$\begin{pmatrix}\theta_1\\\theta_2\\\theta_3\end{pmatrix}=\begin{pmatrix}1.8662\\\pi\\\pi/2\end{pmatrix},\quad\begin{pmatrix}1.8662\\-\pi/2\\-\pi/2\end{pmatrix},\quad\begin{pmatrix}\pi/4\\\pi/2\\\pi/2\end{pmatrix}\text{ or }\begin{pmatrix}\pi/4\\\pi\\-\pi/2\end{pmatrix}\text{ radians}$$

5.5 (ii)

$\cos\theta_1$	$\dfrac{-(31+20\sqrt{2})}{(32+25\sqrt{2})}$	$\dfrac{-(31+20\sqrt{2})}{(32+25\sqrt{2})}$	$1/\sqrt{2}$	$1/\sqrt{2}$
$\sin\theta_1$	$\dfrac{(123+90\sqrt{2})}{(260+189\sqrt{2})}$	$\dfrac{(123+90\sqrt{2})}{(260+189\sqrt{2})}$	$-1/\sqrt{2}$	$-1/\sqrt{2}$
$\cos\theta_2$	0	$1/\sqrt{2}$	0	$1/\sqrt{2}$
$\sin\theta_2$	-1	$-1/\sqrt{2}$	-1	$-1/\sqrt{2}$
$\cos\theta_3$	$1/\sqrt{2}$	$1/\sqrt{2}$	$1/\sqrt{2}$	$1/\sqrt{2}$
$\sin\theta_3$	$1/\sqrt{2}$	$-1/\sqrt{2}$	$1/\sqrt{2}$	$-1/\sqrt{2}$

The four postures are thus given by:-

$$\begin{pmatrix}\theta_1\\\theta_2\\\theta_3\end{pmatrix}=\begin{pmatrix}-2.6470\\-\pi/2\\\pi/4\end{pmatrix},\quad\begin{pmatrix}-2.6470\\-\pi/4\\-\pi/4\end{pmatrix},\quad\begin{pmatrix}-\pi/4\\-\pi/2\\\pi/4\end{pmatrix}\text{ or }\begin{pmatrix}-\pi/4\\-\pi/4\\-\pi/4\end{pmatrix}\text{ radians}$$

5.6   The inverse kinematics of the Stanford manipulator: in the home position the wrist centre is located at $(l_1,0,0)$. The general position of this point is given by the forward kinematics as:-

$$\begin{pmatrix}x_c\\y_c\\z_c\\1\end{pmatrix}=\mathbf{A}_1(\theta_1)\mathbf{A}_2(\theta_2)\mathbf{A}_3(d_3)\begin{pmatrix}l_1\\0\\0\\1\end{pmatrix}=\begin{pmatrix}l_1\cos\theta_1+d_3\sin\theta_1\sin\theta_2\\l_1\sin\theta_1-d_3\cos\theta_1\sin\theta_2\\d_3\cos\theta_2\\1\end{pmatrix}$$

This gives three equations:-

$$x_c = l_1\cos\theta_1+d_3\sin\theta_1\sin\theta_2 \tag{A}$$
$$y_c = l_1\sin\theta_1-d_3\cos\theta_1\sin\theta_2 \tag{B}$$
$$z_c = d_3\cos\theta_2 \tag{C}$$

From these we may form the following equations:-

$$(A)\cos\theta_1 + (B)\sin\theta_1 \equiv x_c\cos\theta_1 + y_c\sin\theta_1 = l_1 \qquad (D)$$
$$(A)\sin\theta_1 - (B)\cos\theta_1 \equiv x_c\sin\theta_1 - y_c\cos\theta_1 = d_3\sin\theta_2 \qquad (E)$$

Squaring and adding these equations we obtain:-

$$(C)^2 + (D)^2 + (E)^2 \equiv x_c^2 + y_c^2 + z_c^2 = l_1^2 + d_3^2$$

This yields an expression for the length of the third joint:-

$$d_3 = \pm\sqrt{x_c^2 + y_c^2 + z_c^2 - l_1^2}$$

Using this result and (C) we have:-

$$\cos\theta_2 = z_c/d_3$$

and hence:-

$$\sin\theta_2 = \pm\sqrt{1 - z_c^2/d_3^2}$$

Returning to the equations (D) and (E) we can find the sine and cosine of θ_1:-

$$x_c\cos\theta_1 + y_c\sin\theta_1 = l_1$$
$$x_c\sin\theta_1 - y_c\cos\theta_1 = \pm\sqrt{d_3^2 - z_c^2}$$

Solving these two equations gives:-

$$\cos\theta_1 = \frac{1}{(x_c^2 - y_c^2)}\{l_1 x_c \mp y_c\sqrt{d_3^2 - z_c^2}\}$$

$$\sin\theta_1 = \frac{1}{(x_c^2 - y_c^2)}\{l_1 y_c \pm x_c\sqrt{d_3^2 - z_c^2}\}$$

The two sign ambiguities introduced produced four postures.

5.7   The lengths are given by:-

$$d_{a1} = |\mathbf{R}\,\mathbf{a} + \mathbf{t} - \mathbf{p}_1|$$
$$d_{b1} = |\mathbf{R}\,\mathbf{b} + \mathbf{t} - \mathbf{p}_1|$$
$$d_{b2} = |\mathbf{R}\,\mathbf{b} + \mathbf{t} - \mathbf{p}_2|$$

Now with $\mathbf{a} = (0, 1)$, $\mathbf{b} = (1, 1)$ and $\mathbf{p}_1$, $\mathbf{p}_2$ as given. We get:-

$$d_{a1} = \sqrt{(t_x - \sin\theta)^2 + (t_y + \cos\theta)^2}$$
$$d_{b1} = \sqrt{(t_x + \cos\theta - \sin\theta)^2 + (t_y + \cos\theta + \sin\theta)^2}$$
$$d_{b2} = \sqrt{(1 + t_x + \cos\theta - \sin\theta)^2 + (t_y + \cos\theta + \sin\theta)^2}$$

Chapter 6

6.1 (i) Choose the home position to be where the first joint lies along the z-axis and the rest of the arm is stretched along the x-axis. The forward kinematics is then:-

$$
\begin{aligned}
x_p &= (l_1 + l_2 \cos\theta_2 + l_3 \cos(\theta_2 + \theta_3))\cos\theta_1 \\
y_p &= (l_1 + l_2 \cos\theta_2 + l_3 \cos(\theta_2 + \theta_3))\sin\theta_1 \\
z_p &= l_2 \sin\theta_2 + l_3 \sin(\theta_2 + \theta_3)
\end{aligned}
$$

6.1 (ii) The jacobian matrix is given by:-

$$
\mathbf{J} = \begin{pmatrix}
-R\sin\theta_1 & \frac{\partial R}{\partial\theta_2}\cos\theta_1 & \frac{\partial R}{\partial\theta_3}\cos\theta_1 \\
R\cos\theta_1 & \frac{\partial R}{\partial\theta_2}\sin\theta_1 & \frac{\partial R}{\partial\theta_3}\sin\theta_1 \\
0 & \frac{\partial z_p}{\partial\theta_2} & \frac{\partial z_p}{\partial\theta_3}
\end{pmatrix}
$$

where $R = (l_1 + l_2 \cos\theta_2 + l_3 \cos(\theta_2 + \theta_3))$ is the distance from the point to the z-axis.

6.1 (iii) Expanding the determinant about the first column gives:-

$$
\det \mathbf{J} = R\det \begin{pmatrix}
\frac{\partial R}{\partial\theta_2} & \frac{\partial R}{\partial\theta_3} \\
\frac{\partial z_p}{\partial\theta_2} & \frac{\partial z_p}{\partial\theta_3}
\end{pmatrix} = -Rl_2l_3 \sin\theta_3
$$

If $l_1 > l_2 + l_3$ then $R \neq 0$ so the singularities are given by $\sin\theta_3 = 0$. Since this does not give a restriction on the other two joints, the singularities define two concentric tori, one given by $\theta_3 = 0$, when the last joint is at full stretch. For the other, $\theta_3 = \pi$, so the last joint is doubled back on itself.

6.1 (iv) From the forward kinematics we have:-

$$
\pm\sqrt{x_p^2 + y_p^2} = l_1 + l_2 \cos\theta_2 + l_3 \cos(\theta_2 + \theta_3)
$$

If $l_1 > l_2 + l_3$, then only the positive sign is possible. The problem then reduces to the two joint planar manipulator of section 5.1.

$$
\begin{aligned}
\sqrt{x_p^2 + y_p^2} - l_1 &= l_2 \cos\theta_2 + l_3 \cos(\theta_2 + \theta_3) \\
z_p &= l_2 \sin\theta_2 + l_3 \sin(\theta_2 + \theta_3)
\end{aligned}
$$

Hence, in general, there are two postures.

6.1(v) However, if $l_1 < l_2 + l_3$ it is possible that there are points with four postures. If $l_1 + l_2 \cos\theta_2 + l_3 \cos(\theta_2 + \theta_3)$ can be negative, then x_p and y_p will remain the same if the signs of $\cos\theta_1$ and $\sin\theta_1$ are reversed. This means that θ_1 changes by π. We get two postures, as before, with $\theta_1 = \tan^{-1}(y_p/x_p)$ and then another two with $\theta_1 = \pi + \tan^{-1}(y_p/x_p)$. However, not all points will be reachable in both these configurations.

Now it is possible that $R = 0$, in which case the point $\mathbf{p}$ is on the axis of the first joint. The forward kinematics is then:-

$$x_p = 0, \qquad y_p = 0, \qquad z_p = l_2 \sin \theta_2 + l_3 \sin(\theta_2 + \theta_3)$$

Since θ_1 does not appear in these equations it can take any value.

6.2   See exercise 5.7. Elements of the jacobian are given by:-

$$\frac{\partial d_{a1}}{\partial t_x} = (t_x - \sin \theta)/d_{a1}$$

$$\frac{\partial d_{a1}}{\partial t_y} = (t_y + \cos \theta)/d_{a1}$$

$$\frac{\partial d_{a1}}{\partial \theta} = -(t_x \cos \theta + t_y \sin \theta)/d_{a1}$$

$$\frac{\partial d_{a2}}{\partial t_x} = (t_x + \cos \theta - \sin \theta)/d_{a2}$$

$$\frac{\partial d_{a2}}{\partial t_y} = (t_y + \cos \theta + \sin \theta)/d_{a2}$$

$$\frac{\partial d_{a2}}{\partial \theta} = -(t_x \cos \theta + t_x \sin \theta - t_y \cos \theta + t_y \sin \theta)/d_{a2}$$

$$\frac{\partial d_{b2}}{\partial t_x} = (1 + t_x + \cos \theta - \sin \theta)/d_{b2}$$

$$\frac{\partial d_{a2}}{\partial t_y} = (t_y + \cos \theta + \sin \theta)/d_{b2}$$

$$\frac{\partial d_{a2}}{\partial \theta} = -(\cos \theta + \sin \theta + t_x \cos \theta + t_x \sin \theta - t_y \cos \theta + t_y \sin \theta)/d_{b2}$$

6.3 (i) The position of a general point is given by:-

$$\begin{pmatrix} x(t) \\ y(t) \\ 1 \end{pmatrix} = \begin{pmatrix} 1 & 0 & \alpha d(t) \\ 0 & 1 & \beta d(t) \\ 0 & 0 & 1 \end{pmatrix} \begin{pmatrix} x(0) \\ y(0) \\ 1 \end{pmatrix}$$

The derivative of this equation when $t = 0$ is:-

$$\begin{pmatrix} \dot{x}(0) \\ \dot{y}(0) \\ 0 \end{pmatrix} = \begin{pmatrix} 0 & 0 & \alpha \dot{d}(0) \\ 0 & 0 & \beta \dot{d}(0) \\ 0 & 0 & 0 \end{pmatrix} \begin{pmatrix} x(0) \\ y(0) \\ 1 \end{pmatrix}$$

Hence we can write:-

$$\begin{pmatrix} \dot{x} \\ \dot{y} \end{pmatrix} = \begin{pmatrix} \alpha \\ \beta \end{pmatrix} \dot{d}$$

So the corresponding column in the jacobian is given by $\begin{pmatrix} \alpha \\ \beta \end{pmatrix}$.

6.3 (ii) The position of the point Q is given by:-

$$\begin{aligned} x_Q &= (1+d)\cos\theta \\ y_Q &= (1+d)\sin\theta \end{aligned}$$

The jacobian is thus given by:-

$$\mathbf{J} = \begin{pmatrix} -(1+d)\sin\theta & \cos\theta \\ (1+d)\cos\theta & \sin\theta \end{pmatrix}$$

6.4   The general positions of the joint vector are given by:-

$$\mathbf{j}_1 = \begin{pmatrix} 0 \\ 1 \\ 0 \end{pmatrix}$$

$$\mathbf{j}_2 = \mathbf{R}(\theta_1,\mathbf{j}) \begin{pmatrix} 1/\sqrt{2} \\ 1/\sqrt{2} \\ 0 \end{pmatrix} = \begin{pmatrix} \frac{1}{\sqrt{2}}\cos\theta_1 \\ \frac{1}{\sqrt{2}} \\ -\frac{1}{\sqrt{2}}\sin\theta_1 \end{pmatrix}$$

$$\mathbf{j}_3 = \mathbf{R}(\theta_1,\mathbf{j})\mathbf{R}(\theta_2,\hat{\mathbf{v}}) \begin{pmatrix} 1 \\ 0 \\ 0 \end{pmatrix} = \begin{pmatrix} \frac{1}{2}\cos\theta_1(1+\cos\theta_2) - \frac{1}{\sqrt{2}}\sin\theta_1\sin\theta_2 \\ \frac{1}{2}(1-\cos\theta_2) \\ -\frac{1}{2}\sin\theta_1(1+\cos\theta_2) - \frac{1}{\sqrt{2}}\cos\theta_1\sin\theta_2 \end{pmatrix}$$

see exercise 4.2. So the jacobian matrix is:-

$$\mathbf{J} = \begin{pmatrix} 0 & \frac{1}{\sqrt{2}}\cos\theta_1 & \frac{1}{2}\cos\theta_1(1+\cos\theta_2) - \frac{1}{\sqrt{2}}\sin\theta_1\sin\theta_2 \\ 1 & \frac{1}{\sqrt{2}} & \frac{1}{2}(1-\cos\theta_2) \\ 0 & -\frac{1}{\sqrt{2}}\sin\theta_1 & -\frac{1}{2}\sin\theta_1(1+\cos\theta_2) - \frac{1}{\sqrt{2}}\cos\theta_1\sin\theta_2 \end{pmatrix}$$

6.5   Differentiating $\mathbf{R}\mathbf{R}^T = \mathbf{I}$ gives:

$$\dot{\mathbf{R}}\mathbf{R}^T + \mathbf{R}\dot{\mathbf{R}}^T = 0$$

Using the fact that $\mathbf{R}(0) = \mathbf{I}$ means that:-

$$\dot{\mathbf{R}}(0) + \dot{\mathbf{R}}^T(0) = 0$$

In other words $\dot{\mathbf{R}}(0)$ is antisymmetric.

6.6   Under the rigid transformation the direction of the axis is rotated:-

$$\hat{\mathbf{v}} \longmapsto \mathbf{R}\hat{\mathbf{v}}$$

but any point on the axis is rotated and translated:-

$$\mathbf{u} \longmapsto \mathbf{R}\,\mathbf{u} + \mathbf{t}$$

The vector product thus becomes:-

$$\mathbf{u} \wedge \hat{\mathbf{v}} \longmapsto (\mathbf{R}\,\mathbf{u} + \mathbf{t}) \wedge \mathbf{R}\,\hat{\mathbf{v}} = \mathbf{R}\,(\mathbf{u} \wedge \hat{\mathbf{v}}) + \mathbf{t} \wedge \mathbf{R}\,\hat{\mathbf{v}}$$

Now, since we may write the operator $\mathbf{t}\wedge$ as an antisymmetric matrix $\mathbf{T}$, see exercise 2.7, the transformation can be summarized as:-

$$\begin{pmatrix} \hat{\mathbf{v}} \\ \mathbf{u} \wedge \hat{\mathbf{v}} \end{pmatrix} \longmapsto \begin{pmatrix} \mathbf{R} & 0 \\ \mathbf{TR} & \mathbf{R} \end{pmatrix} \begin{pmatrix} \hat{\mathbf{v}} \\ \mathbf{u} \wedge \hat{\mathbf{v}} \end{pmatrix} = \begin{pmatrix} \mathbf{R}\,\hat{\mathbf{v}} \\ (\mathbf{R}\,\mathbf{u} + \mathbf{t}) \wedge \mathbf{R}\,\hat{\mathbf{v}} \end{pmatrix}$$

Chapter 7

7.1 (i) The inverse kinematics gives:-

$$\boldsymbol{\theta}(0) = \begin{pmatrix} -0.869 \\ \pi/6 \end{pmatrix}, \qquad \boldsymbol{\theta}(1) = \begin{pmatrix} \pi/6 \\ \pi/6 \end{pmatrix}$$

See also exercise 5.1. So the linear approximation is:-

$$\boldsymbol{\theta}_l(t) = \boldsymbol{\theta}(0)(1 - t) + \boldsymbol{\theta}(1)t = \begin{pmatrix} 1.391t - 0.869 \\ 0.524 \end{pmatrix}$$

7.1 (ii) The inverse kinematics for the mid-point is given by:-

$$\boldsymbol{\theta}(\tfrac{1}{2}) = \begin{pmatrix} -0.465 \\ 1.575 \end{pmatrix}$$

So the quadratic approximation is given by:-

$$\begin{aligned}
\boldsymbol{\theta}_q(t) &= 2(1 - t)(\tfrac{1}{2} - t)\boldsymbol{\theta}(0) + (1 - t)t\boldsymbol{\theta}(\tfrac{1}{2}) + 2(\tfrac{1}{2} - t)t\boldsymbol{\theta}(1) \\
&= \begin{pmatrix} 1.513t^2 + 1.618t - 0.869 \\ -1.575t^2 - 0.521t + 0.524 \end{pmatrix}
\end{aligned}$$

7.2   The end points of the path are:-

$$\mathbf{K}(0) = \begin{pmatrix} \frac{1}{\sqrt{2}} & 0 & \frac{1}{\sqrt{2}} \\ 0 & 1 & 0 \\ -\frac{1}{\sqrt{2}} & 0 & \frac{1}{\sqrt{2}} \end{pmatrix}, \qquad \text{and} \qquad \mathbf{K}(1) = \begin{pmatrix} \frac{1}{\sqrt{2}} & 0 & \frac{1}{\sqrt{2}} \\ 0 & -1 & 0 \\ \frac{1}{\sqrt{2}} & 0 & -\frac{1}{\sqrt{2}} \end{pmatrix}$$

We have that:-

$$\boldsymbol{\theta}(0) = \begin{pmatrix} 0 \\ \pi/4 \\ 0 \end{pmatrix}$$

The inverse kinematics gives the joint angles at the other end as:-

$$\boldsymbol{\theta}(1) = \begin{pmatrix} 0 \\ 3\pi/4 \\ \pi \end{pmatrix} \quad \text{or} \quad \begin{pmatrix} \pi \\ -3\pi/4 \\ 0 \end{pmatrix}$$

Using the second of the solutions would take us through the singularity at $\theta_2 = 0$, so we use the first answer:-

$$\boldsymbol{\theta}(t) = (1-t)\begin{pmatrix} 0 \\ \pi/4 \\ 0 \end{pmatrix} + t\begin{pmatrix} 0 \\ 3\pi/4 \\ \pi \end{pmatrix} = \begin{pmatrix} 0 \\ \pi/4 + t\pi/2 \\ \pi t \end{pmatrix}$$

The mid-point is given by:-

$$\mathbf{K}(\tfrac{1}{2}) = \begin{pmatrix} \frac{1}{\sqrt{2}} & 0 & \frac{1}{\sqrt{2}} \\ \frac{1}{\sqrt{2}} & 0 & -\frac{1}{\sqrt{2}} \\ 0 & 1 & 0 \end{pmatrix}, \quad \text{with joint angles} \quad \boldsymbol{\theta}(\tfrac{1}{2}) = \begin{pmatrix} -\pi/4 \\ \pi/2 \\ \pi/2 \end{pmatrix}$$

The quadratic approximation is thus:-

$$\begin{aligned}
\boldsymbol{\theta}_q(t) &= (1 - 3t + 2t^2)\begin{pmatrix} 0 \\ \pi/4 \\ 0 \end{pmatrix} + (t - t^2)\begin{pmatrix} -\pi/4 \\ \pi/2 \\ \pi/2 \end{pmatrix} + (2t^2 - t)\begin{pmatrix} 0 \\ 3\pi/4 \\ \pi \end{pmatrix} \\
&= \begin{pmatrix} t^2\pi/4 - t\pi/4 \\ 3t^2\pi/2 + t\pi/2 + \pi/4 \\ t^2\pi/2 \end{pmatrix}
\end{aligned}$$

7.3 The approximating polynomial is a quartic:-

$$\boldsymbol{\theta}_{(4)}(t) = \mathbf{a}t^4 + \mathbf{b}t^3 + \mathbf{c}t^2 + \mathbf{d}t + \mathbf{e}$$

The points and derivatives we must match are given by:-

$$\begin{aligned}
\boldsymbol{\theta}_{(4)}(0) &= \mathbf{e} \\
\boldsymbol{\theta}_{(4)}(\tfrac{1}{2}) &= \mathbf{a}/16 + \mathbf{b}/8 + \mathbf{c}/4 + \mathbf{d}/2 + \mathbf{e} \\
\boldsymbol{\theta}_{(4)}(1) &= \mathbf{a} + \mathbf{b} + \mathbf{c} + \mathbf{d} + \mathbf{e} \\
\dot{\boldsymbol{\theta}}_{(4)}(0) &= \mathbf{d} \\
\dot{\boldsymbol{\theta}}_{(4)}(1) &= 4\mathbf{a} + 3\mathbf{b} + 2\mathbf{c} + \mathbf{d}
\end{aligned}$$

These values must be the same as those for the desired path. This gives five linear equations in the five coefficients. The solution can be found by Gaussian elimination:-

$$a = -8\theta(0) + 16\theta(\tfrac{1}{2}) - 8\theta(1) - 2\dot{\theta}(0) + 2\dot{\theta}(1)$$

$$b = 18\theta(0) - 32\theta(\tfrac{1}{2}) + 14\theta(1) + 5\dot{\theta}(0) - 3\dot{\theta}(1)$$

$$c = 11\theta(0) + 16\theta(\tfrac{1}{2}) - 5\theta(1) - 4\dot{\theta}(0) + \dot{\theta}(1)$$

$$d = \dot{\theta}(0)$$

$$e = \theta(0)$$

These equations are given in terms of θ's, which are given, but could be found using the inverse kinematics, and $\dot{\theta}$'s. However, it is the velocity of the path in work space that we must match; this is given by $\mathbf{v} = \mathbf{p}_c(1) - \mathbf{p}_c(0)$. To find the derivatives of the joint angles we must use the jacobian:-

$$\dot{\theta}(0) = \mathbf{J}^{-1}(0)\mathbf{v}, \qquad \dot{\theta}(1) = \mathbf{J}^{-1}(1)\mathbf{v}$$

After some calculation we have that:-

$$\mathbf{J}(0) = \begin{pmatrix} 3/\sqrt{2} & -4/\sqrt{2} - 4/\sqrt{2} & \\ 5/\sqrt{2} & 4/\sqrt{2} & -4/\sqrt{2} \\ 0 & -4 & 0 \end{pmatrix},$$

$$\mathbf{J}(1) = \begin{pmatrix} (2 + 5/\sqrt{2}) & -2 & 0 \\ -(2 + 3/\sqrt{2}) & -2 & 0 \\ 0 & -(4 + 2\sqrt{2}) & -4 \end{pmatrix}$$

and the velocity is:-

$$\mathbf{v} = \begin{pmatrix} -2 - 4\sqrt{2} \\ -2 - \sqrt{2} \\ -4 + 2\sqrt{2} \end{pmatrix}$$

Rather than invert the matrices, it is simpler to solve the linear equations. This gives:-

$$\dot{\theta}(0) = \begin{pmatrix} -1 + 2\sqrt{2} \\ 1 - 1/\sqrt{2} \\ 1/4 + 5/\sqrt{2} \end{pmatrix}, \qquad \dot{\theta}(1) = \begin{pmatrix} -3/2 + 3/2\sqrt{2} \\ 11/8 + 7/4\sqrt{2} \\ -5/4 - 33/8\sqrt{2} \end{pmatrix}$$

Putting all this together gives:-

$$\theta_{(4)}(t) = \begin{pmatrix} -9.97 \\ -3.89 \\ -17.62 \end{pmatrix} t^4 + \begin{pmatrix} 24.48 \\ 12.25 \\ 27.67 \end{pmatrix} t^3 + \begin{pmatrix} -17.90 \\ -9.45 \\ -14.63 \end{pmatrix} t^2 + \begin{pmatrix} 1.83 \\ 0.29 \\ 3.79 \end{pmatrix} t + \begin{pmatrix} 0.79 \\ 1.57 \\ 1.57 \end{pmatrix}$$

7.4   From the three points we have the three linear equations:-

$$b = \theta(0)$$
$$a - \theta(\tfrac{1}{2})c = 2\theta(\tfrac{1}{2}) - 2\theta(0)$$
$$a - \theta(1)c = \theta(1) - \theta(0)$$

These are the same for each component, so we have omitted the subscript. Solving these equations gives:-

$$a_i = \frac{\theta_i(0)\theta_i(\tfrac{1}{2}) - 2\theta_i(0)\theta_i(1) + \theta_i(1)\theta_i(\tfrac{1}{2})}{\theta_i(1) - \theta_i(\tfrac{1}{2})}$$

$$b_i = \theta_i(0)$$

$$c_i = \frac{2\theta_i(\tfrac{1}{2}) - \theta_i(0) - \theta_i(1)}{\theta_i(1) - \theta_i(\tfrac{1}{2})}$$

The coefficients are thus:-

$$a_1 = -2.769, \qquad a_2 = 28.399, \qquad a_3 = -5.486$$
$$b_1 = 0.785, \qquad b_2 = 1.571, \qquad b_3 = 1.571$$
$$c_1 = 1.525, \qquad c_2 = -7.594, \qquad c_3 = -1.431$$

Note that care must be taken with rational approximations, to avoid the denominator vanishing.

Chapter 8

8.1   The wrenches acting on the link are:-

$$\tau_1 = \begin{pmatrix} 0 \\ 0 \\ 1 \\ 0 \\ 0 \\ 0 \end{pmatrix}, \qquad \mathcal{G} = \begin{pmatrix} 0 \\ 0 \\ 0 \\ 0 \\ 0 \\ -1 \end{pmatrix}, \qquad \tau_2 = \begin{pmatrix} -2 \\ 0 \\ 0 \\ 0 \\ 0 \\ 0 \end{pmatrix}$$

Thus the total wrench is given by:-

$$\mathcal{W} = \begin{pmatrix} -2 \\ 0 \\ 1 \\ 0 \\ 0 \\ -1 \end{pmatrix}$$

The pitch p is given by:-

$$p = \begin{pmatrix} -2 \\ 0 \\ 1 \end{pmatrix} \cdot \begin{pmatrix} 0 \\ 0 \\ -1 \end{pmatrix} / \begin{pmatrix} 0 \\ 0 \\ 1 \end{pmatrix} \cdot \begin{pmatrix} 0 \\ 0 \\ -1 \end{pmatrix} = -1/1 = -1$$

The unit vector in the direction of the axis is:-

$$\hat{\mathbf{F}} = \begin{pmatrix} 0 \\ 0 \\ -1 \end{pmatrix}$$

To find a point $\mathbf{r}$ on the axis use the fact that:-

$$\mathbf{r} \wedge \hat{\mathbf{F}} + p\hat{\mathbf{F}} = \begin{pmatrix} -2 \\ 0 \\ 1 \end{pmatrix}$$

$$\mathbf{r} \wedge \begin{pmatrix} 0 \\ 0 \\ -1 \end{pmatrix} = \begin{pmatrix} -2 \\ 0 \\ 0 \end{pmatrix}$$

$$\begin{pmatrix} -r_y \\ r_x \\ 0 \end{pmatrix} = \begin{pmatrix} -2 \\ 0 \\ 0 \end{pmatrix}$$

This has the general solution:-

$$\mathbf{r} = \begin{pmatrix} 0 \\ 2 \\ 0 \end{pmatrix} + \lambda \begin{pmatrix} 0 \\ 0 \\ 1 \end{pmatrix}$$

Here λ is an arbitrary parameter. The point on the axis whose position vector is perpendicular to the axis, $\mathbf{r} \cdot \hat{\mathbf{F}} = 0$, is given by the above vector with $\lambda = 0$.

8.2 First set up a convenient co-ordinate system; the easiest seems have the origin at the centre of the second joint, with the x-axis along the second joint axis, the y-axis along the centre line of the second link and the z axis along the axis of the first joint. With these co-ordinates the wrenches acting on the second link are:-

$$\begin{pmatrix} \mathbf{i} \\ \mathbf{0} \end{pmatrix} + \begin{pmatrix} -\mathbf{i} \\ -\mathbf{k} \end{pmatrix} + \begin{pmatrix} -2\mathbf{k} \\ \mathbf{i} \end{pmatrix} + \mathcal{R}_2 = 0$$

So the reaction wrench at the second joint is:-

$$\mathcal{R}_2 = \begin{pmatrix} 2\mathbf{k} \\ -\mathbf{i} + \mathbf{k} \end{pmatrix}$$

For the first link the wrenches are:-

$$\begin{pmatrix} \mathbf{0} \\ -\mathbf{k} \end{pmatrix} + \begin{pmatrix} 2\mathbf{k} \\ \mathbf{0} \end{pmatrix} + \begin{pmatrix} -\mathbf{i} \\ \mathbf{0} \end{pmatrix} + \mathcal{R}_1 - \mathcal{R}_2 = 0$$

Hence the reaction wrench on the first joint is:-

$$\mathcal{R}_1 = \begin{pmatrix} \mathbf{i} \\ -\mathbf{i} + 2\mathbf{k} \end{pmatrix}$$

8.3 (i) The unit wrenches have components:-

$$\mathbf{M}_{i\mu} = (\mathbf{p}_i \wedge \mathbf{p}_\mu)/|\mathbf{p}_i - \mathbf{p}_\mu|, \qquad \mathbf{F}_{i\mu} = (\mathbf{p}_i - \mathbf{p}_\mu)/|\mathbf{p}_i - \mathbf{p}_\mu|$$

where $i = 1, 2, 3$ and $\mu = a, b, c$. Multiplying by the given magnitudes gives the six wrenches acting on the movable link:-

$$W_{1a} = \sqrt{\frac{3}{7}} \begin{pmatrix} 0 \\ 0 \\ 0 \\ 0 \\ 2/\sqrt{3} \\ 1 \end{pmatrix}, \quad W_{1b} = -\sqrt{\frac{3}{7}} \begin{pmatrix} 0 \\ 0 \\ 0 \\ 1 \\ -1/\sqrt{3} \\ 1 \end{pmatrix}$$

$$W_{2b} = \sqrt{\frac{3}{7}} \begin{pmatrix} 0 \\ -2 \\ -2/\sqrt{3} \\ -1 \\ -1/\sqrt{3} \\ 1 \end{pmatrix}, \quad W_{2c} = -\sqrt{\frac{3}{7}} \begin{pmatrix} 0 \\ -2 \\ 4/\sqrt{3} \\ 0 \\ 2/\sqrt{3} \\ 1 \end{pmatrix}$$

$$W_{3c} = \sqrt{\frac{3}{7}} \begin{pmatrix} \sqrt{3} \\ -1 \\ -4/\sqrt{3} \\ 1 \\ 1/\sqrt{3} \\ 1 \end{pmatrix}, \quad W_{3a} = -\sqrt{\frac{3}{7}} \begin{pmatrix} \sqrt{3} \\ -1 \\ 2/\sqrt{3} \\ -1 \\ -1/\sqrt{3} \\ 1 \end{pmatrix}$$

The total wrench is thus:-

$$W_{tot} = \sqrt{\frac{3}{7}} \begin{pmatrix} 0 \\ 0 \\ -12/\sqrt{3} \\ 0 \\ 2/\sqrt{3} \\ 0 \end{pmatrix}$$

This has pitch $p = 0$, so it is a pure force in the y-direction, and the point (-6,0,0) is on the axis.

8.3 (ii) To find the magnitude of the forces we must solve:-

$$\sqrt{\frac{3}{7}} \begin{pmatrix} 0 & 0 & 0 & 0 & \sqrt{3} & \sqrt{3} \\ 0 & 0 & -2 & -2 & -1 & -1 \\ 0 & 0 & -2/\sqrt{3} & 4/\sqrt{3} & -4/\sqrt{3} & 2/\sqrt{3} \\ 0 & 1 & -1 & 0 & 1 & -1 \\ 2/\sqrt{3} & -1/\sqrt{3} & -1/\sqrt{3} & 2/\sqrt{3} & 1/\sqrt{3} & -1/\sqrt{3} \\ 1 & 1 & 1 & 1 & 1 & 1 \end{pmatrix} \begin{pmatrix} F_{1a} \\ F_{1b} \\ F_{2b} \\ F_{2c} \\ F_{3c} \\ F_{3a} \end{pmatrix}$$

$$= \begin{pmatrix} 6/\sqrt{7} \\ -6\sqrt{3}/\sqrt{7} \\ 0 \\ 0 \\ 2/\sqrt{7} \\ 6\sqrt{3}/\sqrt{7} \end{pmatrix}$$

Note, the columns of the first matrix are the axes of the forces found in exercise 8.3 (i). The solution is:-

$$F_{1a} = 1, \quad F_{1b} = 1, \quad F_{2b} = 1, \quad F_{2c} = 1, \quad F_{3c} = 1, \quad F_{3a} = 1$$

8.4   The total wrench acting on the movable link is:-

$$\mathcal{F}_{\text{tot}} = \sum_{i=1}^{6} f_i \mathcal{F}_i = \left(\mathcal{F}_1 \middle| \mathcal{F}_2 \middle| \cdots \mathcal{F}_6\right) \begin{pmatrix} f_1 \\ f_2 \\ \vdots \\ f_6 \end{pmatrix}$$

So the work done in moving the link is:-

$$\mathcal{F}_{\text{tot}}^T \Delta \mathbf{x} = (f_1, f_2, \ldots, f_6) \begin{pmatrix} \mathcal{F}_1^T \\ \mathcal{F}_2^T \\ \vdots \\ \mathcal{F}_6^T \end{pmatrix} \Delta \mathbf{x}$$

However, the work done can also be expressed as:-

$$\sum_{i=1}^{6} f_i \delta l_i = (f_1, f_2, \ldots, f_6) \Delta \mathbf{l}$$

where $\Delta \mathbf{l}^T = (\delta l_1, \delta l_2, \ldots, \delta l_6)$. Comparing the two expressions for the work done we have that:-

$$\Delta \mathbf{l} = \begin{pmatrix} \mathcal{F}_1^T \\ \mathcal{F}_2^T \\ \vdots \\ \mathcal{F}_6^T \end{pmatrix} \Delta \mathbf{x}$$

since the calculation is valid for arbitrary f_i's. Hence we have that:-

$$\mathbf{J} = \begin{pmatrix} \mathcal{F}_1^T \\ \mathcal{F}_2^T \\ \vdots \\ \mathcal{F}_6^T \end{pmatrix}$$

because $\Delta \mathbf{x}$ is also arbitrary.

Chapter 9

9.1 (i) Taking $a = 2$, $b = 1$ and $c = 3$, then the inertia matrix with respect to the centre of mass is:-

$$I = \begin{pmatrix} \frac{1}{3}(1+9) & 0 & 0 \\ 0 & \frac{1}{3}(4+9) & 0 \\ 0 & 0 & \frac{1}{3}(1+4) \end{pmatrix} = \begin{pmatrix} \frac{10}{3} & 0 & 0 \\ 0 & \frac{13}{3} & 0 \\ 0 & 0 & \frac{5}{3} \end{pmatrix}$$

Now translating the block to the given position:-

$$t = \begin{pmatrix} 0 \\ 7 \\ 3 \end{pmatrix}, \qquad T = \begin{pmatrix} 0 & -3 & 7 \\ 3 & 0 & 0 \\ -7 & 0 & 0 \end{pmatrix}$$

So we must subtract:-

$$T^2 = \begin{pmatrix} -58 & 0 & 0 \\ 0 & -9 & 21 \\ 0 & 21 & -49 \end{pmatrix}$$

Hence, the inertia matrix is:-

$$I' = I - T^2 = \begin{pmatrix} \frac{184}{3} & 0 & 0 \\ 0 & \frac{40}{3} & -21 \\ 0 & -21 & \frac{152}{3} \end{pmatrix}$$

9.1 (ii) For two bodies we have the integrals:-

$$I_1 \omega = \int_{V_1} \rho x \wedge (\omega \wedge x)\, dvol, \qquad I_2 \omega = \int_{V_2} \rho x \wedge (\omega \wedge x)\, dvol$$

where V_1 and V_2 are the regions occupied by the first and second bodies respectively. Since these regions do not overlap we may add the integrals to give:-

$$\int_{V_1+V_2} \rho x \wedge (\omega \wedge x)\, dvol = \int_{V_1} \rho x \wedge (\omega \wedge x)\, dvol + \int_{V_2} \rho x \wedge (\omega \wedge x)\, dvol$$
$$= (I_1 + I_2)\omega$$

Hence, the combined inertia matrix is the sum of the two component inertia matrices. The two blocks in the example have inertia matrices:-

$$I_1 = \begin{pmatrix} \frac{184}{3} & 0 & 0 \\ 0 & \frac{40}{3} & -21 \\ 0 & -21 & \frac{152}{3} \end{pmatrix} \qquad I_2 = \begin{pmatrix} \frac{40}{3} & 0 & 0 \\ 0 & \frac{8}{3} & -3 \\ 0 & -3 & \frac{40}{3} \end{pmatrix}$$

The total inertia matrix is thus:-

$$I_1 + I_2 = \begin{pmatrix} \frac{224}{3} & 0 & 0 \\ 0 & \frac{48}{3} & -24 \\ 0 & -24 & \frac{192}{3} \end{pmatrix}$$

9.2 (i) Consider a general inertia matrix $\mathbf{N} = \begin{pmatrix} \mathbf{I} & M\mathbf{C} \\ M\mathbf{C}^T & M\mathbf{I} \end{pmatrix}$, now move the origin of co-ordinates to the centre of mass of the body:-

$$\mathbf{N}'' = \begin{pmatrix} \mathbf{I} & \mathbf{C}^T \\ 0 & \mathbf{I} \end{pmatrix} \begin{pmatrix} \mathbf{I} & M\mathbf{C} \\ M\mathbf{C}^T & M\mathbf{I} \end{pmatrix} \begin{pmatrix} \mathbf{I} & 0 \\ \mathbf{C} & \mathbf{I} \end{pmatrix} = \begin{pmatrix} \mathbf{I} - M\mathbf{C}^2 & 0 \\ 0 & M\mathbf{I} \end{pmatrix}$$

since $\mathbf{C}$ is antisymmetric. Now rotate the co-ordinate axes so that in the new co-ordinates $\mathbf{R}^T(\mathbf{I} - M\mathbf{C}^2)\mathbf{R} = \mathbf{D}$ is diagonal; the fact that this is always possible for a symmetric matrix is known as Sylvester's theorem. The diagonal entries are then the eigenvalues of the matrix. These must be positive since the kinetic energy of a rigid body is always positive. The diagonal entries are called the principal moments of inertia.

9.2 (ii) First we can find blocks with any mass. Now suppose that in the co-ordinates where the inertia matrix is diagonal, the entries are d_1, d_2 and d_3. Comparing these with the inertia matrix for the rectangular block we have three equations:-

$$\frac{M}{3}(b^2 + c^2) = d_1$$
$$\frac{M}{3}(a^2 + c^2) = d_2$$
$$\frac{M}{3}(a^2 + b^2) = d_3$$

Inverting these relations gives:-

$$a = \sqrt{\frac{3}{2M}(-d_1 + d_2 + d_3)}$$
$$b = \sqrt{\frac{3}{2M}(d_1 - d_2 + d_3)}$$
$$c = \sqrt{\frac{3}{2M}(d_1 + d_2 - d_3)}$$

This fails if the quantity under one of the square roots is negative. So we can only find an equivalent rectangular block if the body's largest principal moment of inertia is smaller than the sum of the other two principal moments of inertia.

9.3 (i)

$$\mathbf{S} \wedge \mathbf{S} = \begin{pmatrix} \psi \wedge \psi \\ \mathbf{u} \wedge \psi + \psi \wedge \mathbf{u} \end{pmatrix}$$

This is zero by the properties of the vector product for vectors in three dimensions.

9.3 (ii) From the definitions we have:-

$$\{\mathbf{S}_1, \mathcal{W}\}^T \mathbf{S}_2 = \psi_2 \cdot (\psi_1 \wedge \gamma) + \psi_2 \cdot (\mathbf{u}_1 \wedge \mathbf{E}) + \mathbf{u}_2 \cdot (\psi_1 \wedge \mathbf{E})$$
$$\mathcal{W}^T(\mathbf{S}_1 \wedge \mathbf{S}_2) = \gamma \cdot (\psi_1 \wedge \psi_2) + \mathbf{E} \cdot (\mathbf{u}_1 \wedge \psi_2) + \mathbf{E} \cdot (\psi_1 \wedge \mathbf{u}_2)$$

These are equal by the cyclic properties of the scalar triple product of three-vectors.

9.4 From section 9.4, the general equation of motion for a single link is:-

$$\mathcal{W}^T \mathbf{S} = \mathbf{S}^T \mathbf{N} \mathbf{S} \ddot{\theta}$$

Here the wrench is just the force due to gravity and the screw is the joint axis:-

$$\mathcal{W} = \mathcal{G} = \begin{pmatrix} -Mg\mathbf{c} \wedge \mathbf{k} \\ -Mg\mathbf{k} \end{pmatrix}, \qquad \mathbf{S} = \begin{pmatrix} \mathbf{k} \\ p\mathbf{k} \end{pmatrix}$$

so the equation of motion is now:-

$$-Mpg = (\mathbf{I}_{33} + Mp^2)\ddot{\theta}$$

The 3,3 component of the inertia matrix $\mathbf{I}_{33}$ is independent of the joint angle in this case.

9.5 The total kinetic energy is the sum of the kinetic energies of the individual links:-

$$\mathrm{KE} = \frac{1}{2} \sum_{i=1}^{6} \mathbf{V}_i^T \mathbf{N}_i \mathbf{V}_i$$

Now since $\mathbf{V}_i = \sum_{j=1}^{i} \dot{\theta}_j \mathbf{S}_j$, we have:-

$$\frac{\partial \mathbf{V}_i}{\partial \dot{\theta}_j} = \begin{cases} \mathbf{S}_j, & \text{if } j \leq i \\ 0, & \text{if } j > i \end{cases}$$

The inertia matrix depends on the joint angles but not (directly) on their velocities. Hence, we have:-

$$\frac{\partial \mathrm{KE}}{\partial \dot{\theta}_j} = \frac{1}{2} \left(\sum_{i=1}^{j} \mathbf{S}_i^T \mathbf{N}_i \mathbf{V}_i + \sum_{i=1}^{j} \mathbf{V}_i^T \mathbf{N}_i \mathbf{S}_i \right) = \sum_{i=1}^{j} \mathbf{V}_i^T \mathbf{N}_i \mathbf{S}_i$$

using the fact that the inertia matrices are symmetric.

9.6 Including friction terms gives:-

$$\tau_i = \sum_{j=i}^{6} \left[\dot{\mathbf{V}}_j^T \mathbf{N}_j \mathbf{S}_i + \mathbf{V}_j^T \mathbf{N}_j (\mathbf{V}_j \wedge \mathbf{S}_i) - \mathcal{G}_j^T \mathbf{S}_i \right] + \mu_i \dot{\theta}_i \qquad i = 1, 2, \ldots, 6$$

here μ_i is the friction coefficient for the ith joint.

Index